Dynamic Modeling of Diseases and Pests

Bruce Hannon • Matthias Ruth

Dynamic Modeling
of Diseases and Pests

 Springer

Bruce Hannon
University of Illinois
220 Davenport Hall, MC 150
Urbana, IL 61801

Matthias Ruth
University of Maryland
2101 Van Munching Hall
College Park, MD 20782

Additional material to this book can be downloaded from http://extras.springer.com.

ISBN: 978-1-4899-9503-2   ISBN: 978-0-387-09560-8 (eBook)

Contents

Part II Applications

Part III Conclusions

Part I
Introduction

Chapter 1
The Why and How of Dynamic Modeling

1.1 Introduction

Few tasks are nobler than those that improve the length and quality of life of humans and their fellow species. And few tasks are more difficult to accomplish. To be successful, we must assess the vulnerabilities of individuals to attacks on their health and well-being, we must understand the interactions of individuals with each other and their environment, and we must anticipate the likely consequences of all these factors in an ever-changing world—individual vulnerabilities change, new diseases and pests emerge, old ones reappear, new means are developed to detect and combat adverse influences on health and well-being, and new standards for health and quality of life are applied.

There are many drivers behind the spread of diseases and pests. Climate change may create new temperature and precipitation regimes conducive to diseases and pests that would otherwise be irrelevant for particular locations. West Nile fever, malaria and encephalitis, for example, are increasingly of concern to public health officials. Other drivers are related to our use of technology in a globalizing world. For example, ballast water used in ships can carry with it organisms and spread them to ever more far-flung places. Increased travel of people around the globe can promote dispersal of viruses, bacteria, fungi, and other "agents" that affect the health and well-being of species and ecosystems.

Strategies for controlling the spread of diseases and pests, and especially changing the root causes for a spread to occur, have created a multi-billion-dollar industry involving the full gamut of societal defenses—from detection and monitoring, to chemical and pharmaceutical products, to medical care, to ecosystem design and restoration. Sound knowledge of the dynamics of diseases and pests and an understanding of the changing roles and relationships among the drivers and the constraints on their spread are needed to make wise choices among the various intervention options.

This book provides an introduction to dynamic modeling of diseases and pests—the various forms of insult to the health and well-being of species, both human and

B. Hannon and M. Ruth, *Dynamic Modeling of Diseases and Pests*,
Modeling Dynamic Systems,
© Springer Science + Business Media LLC 2009

animal. We draw on insights from biology, epidemiology, and related disciplines to identify key components of, and influences on, human and environmental systems. We use the graphical programming language STELLA to organize these insights into formal models that can be run on a computer; we then use these models to investigate the dynamics of pestilence and explore alternative scenarios for outside intervention into the systems' dynamics. In particular, we look for emergent properties of the model—those results that we did not expect.

We consider this kind of modeling as a subtle craft, an art form that is intended to help us understand the future. And because of the complexity of dynamic systems, the use of formal models and numbers is essential—they help us dispel the complexity of many real-world processes and force us to be specific. Good dynamic modeling is an art. It requires modeling experience that draws upon modeling analogies for the creation of new and useful models.

Modeling dynamic systems is central to our understanding of real-world phenomena. We all create dynamic mental models of the world around us, dissecting our observations into cause and effect. Such mental models enable us, for example, to cross a busy street or hit a baseball successfully. But we are not mentally equipped to go much further. The complexities of social, economic, or ecological systems and their interactions force us to use aids if we want to understand much of anything about them.

With the advent of personal computers and graphical programming, everyone can create more sophisticated models of the phenomena in the world around us. As Heinz Pagels noted in *Dreams of Reason* in 1988, the computer modeling process is to the mind what the telescope and the microscope are to the eye. We can model the macroscopic results of microphenoma, and vice versa. We can simulate the various possible futures of a dynamic process. We can begin to explain and perhaps even to predict.

In order to deal with these phenomena, we abstract from details and attempt to concentrate on the larger picture—a particular set of features of the real world or the structure that underlies the processes that lead to the observed outcomes. Models are such abstractions of reality. Models force us to face the results of the structural and dynamic assumptions that we have made in our abstractions.

The process of model construction can be rather involved. However, it is possible to identify a set of general procedures that are followed frequently. These general procedures are shown in simplified circular form (Figure 1.1).

Models help us understand the dynamics of real-world processes by mimicking with the computer the actual but simplified forces that are assumed to result in a system's behavior. For example, it may be assumed that the number of people contracting a disease is directly proportional to the size of the infected and susceptible populations. In a simple version of this epidemic model, we may abstract away from a variety of factors that impede or stimulate the spread of a disease in addition to factors directly related to the different population sizes and distance. Such an abstraction may leave us with a sufficiently good predictor of the known infection rates, or it may not. If it does not, we reexamine the abstractions, reduce the assumptions, and retest the model for its new predictions. Models help us in the

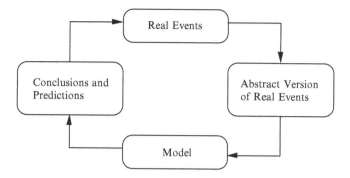

Fig. 1.1

organization of our thoughts and data and in the evaluation of our knowledge about the mechanisms that lead to the system's change.

Some people raise philosophic questions as to why one would want to model a system. As pointed out earlier, we all perform mental models of every dynamic system we face. We also learn that in many cases, those mental models are inadequate. With a formal model at hand—a model that is transparent enough for others to understand and critique, and one that can be run over and over again to reveal its behavior under different assumptions—we can specifically address the needs and rewards of modeling.

Throughout this book, we encounter a variety of nonlinear, time-lagged feedback processes, some with random disturbances that give rise to complex system behavior. Such processes can be found in a large range of systems. The variety of models in the companion books of this series naturally span only a small range—but the insights on which these models are based can (and should) be used to inform the development of models for systems that we do not cover here.

It is our intention to show you how to model, not how to use models, nor how to set up a model for someone else's use. The latter two are certainly worthwhile activities, but we believe that the first step is learning the modeling process. In the following section, we introduce you to the computer language that is used throughout the book. This computer language will be immensely helpful as you develop an understanding of dynamic systems and use that understanding to solve new problems.

1.2 Static, Comparative Static, and Dynamic Models

Most models fit in one of three general classes. The first type consists of models that represent a particular phenomenon at a point of time—these are static models. For example, a map of the United States may depict the location and size of a city or the rate of infection with a particular disease, each in a given year. The second type is the set of comparative static models that compare some phenomena at different points in time. This is like using a series of snapshots to make inferences about the system's path from one point in time to another, without explicitly modeling that process.

Other models describe and analyze the very processes underlying a particular phenomenon. An example of this would be mathematical models that describe host-parasite interactions. Such models could capture population dynamics of hosts and parasites, where at any point in time the population size of one depends on its size during an earlier time period and on the size of the other population. These are dynamic models. Dynamic models try to reflect changes in real or simulated time and take into account that the model components are constantly changing as a result of previous conditions and current influences.

With the advent of easy-to-use computers and software, everyone can build on the existing descriptions of a system and carry them further. While a static or comparative static assessment may be adequate for many purposes, the world itself changes all the time. Models that treat a system in a static or comparative static fashion may be misleading and become obsolete. We can now investigate in great detail and with great precision the system's behavior over time. However, the longer the time frame over which we make such investigations, the more likely it may be that new forces that influence the system's dynamics will come into play. For example, if we model the spread of a disease without considering mutation of disease-causing agents, our model results may be perfectly valid for time frames that are small enough to render mutation irrelevant. In the long run, however, virulence may change as a result of mutation; and thus the dynamics of the disease itself may become different; and thus our results are a less than adequate representation of what we should expect in the long run. Conversely, including the effects of mutation on the spread of a disease may be irrelevant to short-term dynamics.

1.3 Model Complexity and Explanatory Power

What can we do to ensure that a model could produce a sufficiently good predictor of the dynamics of pestilence? If the model does not appear to be a good predictor, we can reexamine it. Did our abstractions eliminate any important factors? Were all our assumptions valid? We can revise the model, based on the answers to these questions. Then, we can test the revised model for its "predictions" of historic events before asking it to project into the real future. We should then have an improved model of the system we are studying. Even better, our understanding of that system will have grown. For example, we may have found out that our model's behavior is virtually unchanged over a relevant time frame of interest, given reasonable assumptions about the mutation of a disease-causing agent. We can now better determine whether we asked the right questions, included all the important factors in that system, and represented those factors properly. We might find that certain of the model parameters have particularly high leverage on important variables while changes in other parameters produce little such change. This knowledge tells us where to put our research effort into refining parameter values.

Elementary to modeling is the idea that a model should be kept simple, even simpler than the cause and effect relationships it studies. Add complexities to the model only when it does not produce the real effects. Remember that models are sketches of real systems and are not designed to show all of a system's many facets. Models aid us in understanding complicated systems by simplifying them.

Models study cause and effect; they are causal. The modeler specifies initial conditions and relations among these elements. The model then describes how each condition will change in response to changes in the others. In the example of host-parasite interactions, a larger population of hosts could provide more opportunities for parasites to grow, reproduce, and spread—which could affect the health of hosts, reduce the growth rate of their population, and thus negatively impact growth of the parasite population.

Key to model design is the decision on system boundaries—what to explicitly model and what to consider as given. Open models are models whose workings are influenced by a large number of outside influences that are not explicitly modeled. For example, your model of the spread of a disease may consider birth and death rates as given, contact rates of infected with susceptible individuals fixed, and any number of other parameters not the actual subject of your model itself. In that case, you have an open model. As you explain more how birth, death, or contact rates change in response to factors that are part of the model, you "internalize" these factors and "close" your model. As you approach a closed model, its complexity is likely to increase as ever more components are influenced by each other.

The initial or starting conditions from which a model runs could be actual measurements (the number of people in a city in a given year) or estimates (the number of people there in four years, given normal birth rate and other specified conditions). Such estimates are designed to reflect the process under study, not to provide precise information about it. Therefore, the estimates could be based on real data or the reasonable guesses of a modeler who has experience with the process. At each step in the modeling process, documentation of the initial conditions, choice of parameters, presumed relationships, and any other assumptions is always necessary, especially when the model is based on the modeler's guesses.

Dynamic models have an interesting interpretation in the world of dynamical statistics. The entire dynamic model in STELLA might be considered as a single regression equation, and it can be used that way in a statistical analysis of the best parameter choice for optimization of a performance measure[1], such as maximizing the effectiveness of a vaccine by choosing the best fraction of the population to be vaccinated. It replaces the arbitrary functional form of the regression equation used in statistical analysis. Thinking of the dynamic model in this way lets one imagine that because more actual system form and information is represented in the

---

[1] Personal paper: Twenty-Five Years (Isolated) Behind Enemy Lines: Do economists and statisticians have anything new to offer Systems Dynamics? George Backus, Policy Assessment Corporation, 14604 West 62nd Place, Arvada, Colorado 80004, George_Backus@Energy2020.com, 2003.

dynamic model, that model will produce more accurate results. This is probably so if the same number of parameters are used in both modeling processes. If this is not a constraint, the statistical dynamics process known as co-integration can produce more accurate results. But one can seldom determine in physical terms what aspect of the co-integration model accounted for its accuracy; we are left with a quandary.

If we wish modeling to do more than simulate a complex process—that is, if we want a model to help us understand how to change and improve a real-world process—we prefer to use dynamic modeling. Statistical analysis is not involved in optimizing the performance measures to complete the dynamic modeling process; it is key to finding the parameters for the inputs to these models. For example, we may use a normal distribution to describe the daily mean temperature to determine the growth rate of bacteria in rivers or lakes throughout the course of a year. Statistical analysis is essential in determining the mean and standard deviation of these temperatures from the temperature record. Thus using statistical analysis, we compress years of daily temperature data into a single equation that is sufficient for our modeling objectives.

1.4 Model Components

Model building begins, of course, with the properly focused question. Then the modelers must decide on the boundaries of the system that contains the question, choose a meaningful time horizon over which to explore system behavior, select an appropriate time increment or "time step" (minutes, days, months, year, decade) for which system change is modeled, and choose an adequate level of detail. But these are verbal descriptions. Sooner or later the modeler must get down to the business of actually building the model. The first step in that process is the identification of the state variables, those variables that will indicate the status of this system through time. These variables carry the knowledge of the system from step to step throughout the run of the model—they are the basis for the calculation of the rest of the variables in the model.

Generally, the two kinds of state variables are conserved and nonconserved. Examples of conserved variables are the population of an island or the water behind a dam. They have no negative meaning. Examples of nonconserved state variables are temperature or price, and they might take on negative values (temperature) or they might not (price).

Control variables are the ones that directly change the state variables. They can increase or decrease the state variables through time. Examples include birth (per time period) or water inflow or outflow (to and from a reservoir).

Transforming or converting variables are sources of information used to change the control variables. A transforming or converting variable might be the result of an equation based on still other transforming variables or parameters. Birth rate, contact rate, or mutation rate are examples of transforming variables.

The components of a model are expected to interact with each other. Such interactions engender feedback processes. *Feedback* describes the process wherein one component of the model initiates changes in other components, and those modifications lead to further changes in the component that set the process in motion. For example, everything else equal, an increase in the size of a population leads to increases in the number of births, which in turn leads to an increase in the size of the population. In positive feedback, the original modification leads to changes that reinforce the component that started the process and typically lead the system away from its initial state. The resulting dynamics are often referred to as "explosive dynamics"—a phrase frequently used in modified forms, such as when we refer to a "population explosion."

Feedback is said to be negative when the modification in a component leads other components to respond by counteracting that change. For example, an increase in a medication dosage may help fight a disease initially, such that lower doses of medication are required later. Negative feedback is often the engine that drives a system toward a steady state. The word *negative* does not imply a value judgment—it merely indicates that feedback tends to negate initial changes.

People from different disciplines perceive the role and strength of feedback processes differently. Economists, for example, are typically preoccupied with market forces that lead to equilibrium in the system. Therefore, the models are dominated by negative feedback mechanisms, such as price increases in response to increased demand. The work of ecologists and biologists, in contrast, is frequently concerned with positive feedback, such as that leading to insect outbreaks or the dominance of hereditary traits in a population.

Most systems contain both positive and negative feedback; these processes are different and vary in strength. For example, as more people are born in a rural area, the population may grow faster (positive feedback). However, as the limits of available arable land are reached by agriculture, the birth rate slows, at first perhaps for psychological reasons but eventually for reasons of starvation (negative feedback).

Nonlinear relationships complicate the study of feedback processes. An example of such a nonlinear relationship would occur when a control variable does not increase in direct proportion to another variable but changes in a nonlinear way. Nonlinear feedback processes can cause systems to exhibit complex—even chaotic—behavior.

A variety of feedback processes engender complex system behavior, and some of these will be covered later in this book. For now, we develop a simple model, which illustrates the concepts of state variables, flows, and feedback processes. Discussion will then return to some "principles of modeling" that will help you to develop the model building process in a set of steps.

Besides feedback, two other real-world properties make purely mental models truly impractical. They are delays and randomness. The response to an action is often delayed in time with the delayed effect arriving sometimes at the most inopportune time or too late for real effect. The time between the recognition of the onset of a serious contagious disease and the implementation of a vaccine program to stave off its worst effects can be great—so great that the negative effects of the vaccine only add to the overall maladies. Randomness exists in all living systems. For example, the effect that air temperature has on the birth rate of mosquitoes carrying West Nile virus can be dramatic. However, trying to predict the critical air tomorrow or over the next two weeks, a long time in the world of the insect, is best modeled using historic temperatures with an added random increment. Then one can examine the range of possible outcomes due to this randomness.

Not only is randomness something that one must deal with in models but it is the source, for good or bad, of the new. Treating the need to consider randomness in models as necessary when one wanted to represent the variations of say weather, or temperature – the kinds of variables for which one could not and need not provide explicit models but whose behavior may appropriately be described by variation around some mean values. But randomness has a more important aspect: The epistemologist Gregory Bateson has remarked that we could not have music and the creation of novel forms unless we had a background of noise, of uncommitted potential in randomness and disorder that awaited selection in the ordering of the creative act. "All that is not information, not redundancy, not form and not restraints – is noise, the only possible source of new patterns."

1.5 Modeling in STELLA

STELLA[2] was chosen as the computer language for this book on Dynamic Modeling of Pestilence because it is a powerful, yet easy-to-learn tool. Readers are expected to familiarize themselves with the many features of the program. Some introductory material is provided in the appendix. Careful reading of the Help File that accompanies the program is advised. Experiment with the STELLA software and become thoroughly familiar with it.

To explore modeling with STELLA, we will develop a basic model of the dynamics of a contagious disease in a human population. Assume that initially, only 10 people are sick and that the contagion rate is 5 percent per day; that is, each day, 5 people become sick for every 100 sick people. For simplicity, assume also that none of the sick people die. How many sick people will we have after 80 days?

In building the model, utilize all four of the graphical "tools" for programming in STELLA. The appendix has a "Quick Help Guide" to the software. The appendix also describes how to install the STELLA software and models of the book. Follow these instructions. Then, double-click on the STELLA icon to open it.

On opening STELLA, you will be faced with the "High-Level Mapping Layer," which is provided to help you develop user interfaces for an existing model as you become more experienced. For now, go to the "Diagram Layer"—that layer in which we actually develop an executable model—and click on the downward-pointing arrow in the upper left-hand corner of the frame (Figure 1.2):

---

[2] All models in this book were made with STELLA complete version 9.

Fig. 1.2

The Diagram Layer displays the following symbols, "building blocks," for stocks, flows, converters, and connectors (information arrows), shown in Figure 1.3:

Fig. 1.3

Click on the globe (Figure 1.4) to access the modeling mode:

Fig. 1.4

In the modeling mode you can specify your model's initial conditions and functional relationships. The following symbol (Figure 1.5) indicates that you are now in the modeling mode:

Fig. 1.5

Begin with the first tool, a stock (rectangle). In this example model, the stock will represent the number of fish in your pond. Click on the rectangle with your mouse, drag it to the center of the screen, and click again. Type in the word SICK. This is what you see (Figure 1.6):

SICK

Fig. 1.6

This is the first state variable in the model. This is where you indicate and document a state or condition of the system. In STELLA, this stock is known as a *reservoir*. In this model, the stock represents the number of sick people. This variable will be updated at every step of time (DT) for which STELLA carries out a calculation and stored in the computer's memory throughout the duration of the model.

The SICK population is a stock, something that can be contained and conserved in the *reservoir*. There are nonconserved state variables in other kinds of models, for example temperature or price.

Inside the rectangle is a question mark. This is to remind you that you need an initial or starting value for all state variables. Double-click on the rectangle. A dialogue box will appear. The box is asking for an initial value. Add the initial value you choose, (in our case, 10) using the keyboard or the mouse and the dialogue keypad. When you have finished, click on OK to close the dialogue box. Note that the question mark has disappeared.

Decide next what factors control (that is, add to or subtract from) the number of sick individuals in the population. Because an earlier assumption was that the SICK in your POPULATION never die, you have one control variable: GETTING SICK. Use the *flow* tool (the right-pointing arrow, second from the left) to represent the control variable, so named because they control the states (variables). Click on the flow symbol, then click on a point about 2 inches to the left of the rectangle (stock) and drag the arrow to SICK, until the stock becomes dashed, and release. Label the circle GETTING SICK. This is what you will have (Figure 1.7):

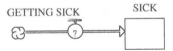

Fig. 1.7

Here, the arrow points only into the stock, which indicates an inflow. But, you can get the arrow to point both ways if you want it to. You do this by double-clicking on the circle in the flow symbol and choosing "Biflow." Biflow enables you to add to the stock if the flow generates a positive number and subtract from the stock if the flow is negative. In this model, of course, the flow GETTING SICK is always positive and newly sick people flow only *into* SICK. The control variable GETTING SICK is a uniflow: new sick per day.

It is a good practice to name the state variables as nouns (e.g., SICK) and the direct controls of the states as verbs (e.g., GETTING SICK). The parameters are most properly named as nouns. This somewhat subtle distinction keeps the flow and stock definitions foremost in the mind of the beginning modeler.

Next you must know how the people in this population become sick—not the biological details, just how to accurately estimate the number of new sick per annum. One way to do this is to look up the contagion rate for the particular disease on the website of the Centers for Disease Control or recent publications in the scientific literature. Suppose that we find the CONTAGION RATE is 5 percent per day, a number that can be represented as a transforming variable. A transforming variable is expressed as a *converter*, the circle that is second from the right in the STELLA toolbox. (So far GETTING SICK is a constant; later the model will allow this rate to vary.) The same clicking and dragging technique that brought the stock to the screen will bring up the circle. Open the converter and enter the number of 0.05 (5/100). Down the side of the dialogue box is an impressive list of built-in functions that are useful for more elaborate models. We'll use some of these built-in functions later on in the book.

At the right of the STELLA toolbox is the *connector* (information arrow). Use the connector to pass on information (about the state, control, or transforming variable) to a circle, to the control (the transforming variable). In this case, you want to pass on information about the CONTAGION RATE to GETTING SICK. Once you draw the information arrow from the transforming variable CONTAGION RATE to the control and from the stock SICK to the control, open the control by double-clicking on it. Recognize that CONTAGION RATE and SICK are two required inputs for the specification of GETTING SICK. Note also that STELLA asks you to specify the control: GETTING SICK = . . . "Place right-hand side of equation here."

Click on CONTAGION RATE, then on the multiplication sign in the dialogue box, and then on SICK to generate the equation

$$\text{GETTING SICK} = \text{CONTAGION RATE} * \text{SICK} \tag{1.1}$$

Click on OK, and the question mark in the control GETTING SICK disappears. Your STELLA diagram should now look like Figure 1.8:

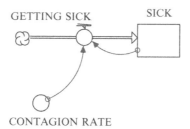

Fig. 1.8

Next, set the temporal (time) parameters of the model. These are DT (the time step over which the stock variables are updated) and the total time length of a model run. Go to the RUN pull-down menu on the menu bar and select Time Specs. A dialogue box will appear in which you can specify, among other things, the length of the simulation, the DT, and the units of time. For this model, choose DT = 1, length of time = 20, and units of time = days.

To display the results of the model, click on the graph icon and drag it to the diagram. If you wanted to, you could display these results in a table by choosing the table icon instead. The STELLA icons for graphs and tables (Figure 1.9) are, respectively,

Fig. 1.9

When you create a new graph pad, it will open automatically. To open a pad that had been created previously, just double-click on it to display the list of stocks, flows, and parameters for the model. Each one can be plotted. Select SICK to be plotted; with the ≫ arrow, add it to the list of selected items. Then set the scale from 0 to 80 and check OK. You can set the scale by clicking once on the variable whose scale you wish to set and then on the arrow next to it. Now you can select the minimum on the graph, and the maximum value will define the highest point on the graph. Rerunning the model under alternative parameter settings will lead to graphs that are plotted over different ranges. Sometimes these are a bit difficult to compare with previous runs, because the scaling automatically changes unless fixed by the modeler.

Would you like to see the results of the model so far? Run the model by selecting RUN from the pull-down menu. You should see the following (Figure 1.10):

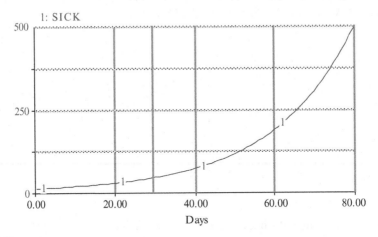

Fig. 1.10

The graph shows exponential growth of the sick population in your population. This is what you should have expected. It is important to state beforehand what results you expect from running a model. Such speculation builds your insight into system behavior and helps you anticipate (and correct) programming errors. When the results do not meet your expectations, something is wrong and you must fix it.

The error may be either in your STELLA program or your understanding of the system that you wish to model—or both.

What do you really have here? How does STELLA determine the time path of the state variable? At the beginning of each time period, starting with time = 0 days (the initial period), STELLA looks at all the components for the required calculations. The values of the state variables will form the basis for these calculations. Only the variable GETTING SICK depends on the state variable SICK. To estimate the value of GETTING SICK after the first time period, STELLA multiplies 0.05 by the value SICK (at time = 0) or 10 (provided by the information arrows) to arrive at 0.5. From time = 1 to time = 2, the next DT, STELLA repeats the process and continues through the length of the model. When you plot your model results in a table, you find that, for this simple model, STELLA calculates fractions of SICK from time = 1 onward. This problem is easy to solve; for example, by having STELLA round the calculated number of SICK—there is a built-in function that can do that—or just by reinterpreting the population size as "thousands of SICK."

This process of calculating stocks from flows highlights the important role played by the state variable. The computer carries that information—and only that information—from one DT to the next, which is why it is defined as the variable that represents the *condition* of the system.

You can drill down in the STELLA model to see the parameters and equations that you have specified and how STELLA makes use of them. Click on the downward-pointing arrow at the left of your STELLA diagram.

The equations and parameters of your models are listed here. The model equations are also listed at the end of each chapter of this book so you can more easily recreate the models. Note how the SICK population in time period t is calculated from the population one small time step, DT, earlier and all the flows that occurred during that DT.

The model of the SICK population dynamics is simple. So simple, in fact, that it could be solved with pencil and paper, using analytic or symbolic techniques. The model is also linear and unrealistic. Next, add a dimension of reality—and explore some of STELLA's flexibility. This may be justified by the observation that, as populations get large, mechanisms set in that influence the rate of GETTING SICK.

To account for feedback between the size of the SICK population and its rate of GETTING SICK, an information arrow is needed to connect SICK with CONTAGION RATE. The connection will cause a question mark to appear in the symbol for CONTAGION RATE. The previous specification is no longer correct; it now requires SICK as an input. We will make this input indirectly. First we construct another converter called AWARENESS LEVEL. This level is a function of the number of SICK, let's say, equal to 0.1 times the number of SICK. This AWARENESS LEVEL then affects the CONTAGION RATE, the greater the awareness of the disease as determined by the sheer number of SICK, the lower the CONTAGION RATE (Figure 1.11).

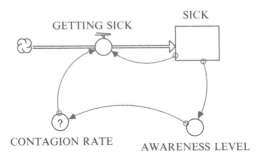

Fig. 1.11

Open CONTAGION RATE. Click on the required input AWARENESS LEVEL. The relationship between AWARENESS LEVEL and CONTAGION RATE must be specified in mathematical terms, or at least, you must make an educated guess about that relationship. An educated guess about the relationship between two variables can be expressed by plotting a graph that reflects the anticipated effect one variable (AWARENESS LEVEL) will have on another (CONTAGION RATE). The feature used for this is called a *graphical function*.

To use a graph to delineate the extended relationship between CONTAGION RATE and AWARENESS LEVEL, open CONTAGION RATE and enter AWARENESS LEVEL as the graph driver. Click on "Become Graphical Function" and set the limits on the AWARENESS LEVEL at 0 and 10. Set the corresponding limits on the CONTAGION RATE at 0 and 0.10, to represent a change in this rate when the AWARENESS LEVEL is between 0 and 10. (These are arbitrary numbers for a made-up model.) Finally, use the mouse arrow to draw a curve from the maximum CONTAGION RATE an AWARENESS LEVEL of 0 to the point of 0 birth rate an AWARENESS LEVEL of 10.

Suppose a survey of the AWARENESS LEVEL and CONTAGION RATE were taken at three points in time. The curve you just drew goes through all three points. You can assume that, if a census had been taken at other times, it would show a gradual transition through all the points (Figure 1.12). This sketch is good enough for now. Click on OK.

Before you run the model again, consider what the results will be. Think of the graph for SICK through time. Generally, it should rise, but not in a straight line. At first the rise should be steep: the initial population is only 10, so the initial CONTAGION RATE should be high. Later it will slow down. Then, the population should level off at 100 (10 times the maximum AWARENESS LEVEL), when SICK would

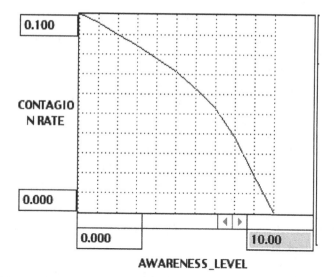

Fig. 1.12

be so great that new contagion tends to cease. Run the model. Indeed, the results (Figure 1.13) are consistent with our expectation—and so they should be!

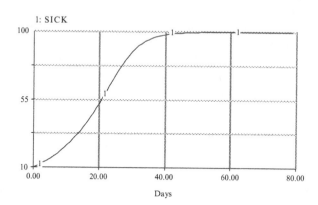

Fig. 1.13

This problem has no analytic solution, only a numerical one. You can continue to study the sensitivity of the answer to changes in the graph and the size of DT. You are not limited to a DT of 1. Generally speaking, a smaller DT leads to more accurate numerical calculations for updating state variables and, therefore, a more accurate answer. Choose "Time Specs" from the RUN menu to change the DT. Change the

DT to reflect ever-smaller periods until the change in the critical variable is within measuring tolerances. The best guide in the determination of the proper modeling time step (DT) is the "half rule." Run the model with what appears to be an appropriate time step, then halve the DT and run the model again, comparing the two results of important model variables. If these results are judged to be sufficiently close, the first DT is adequate. One might try to increase the DT if possible to make the same comparison. The general idea is to set the DT to be significantly smaller than the fastest time constant in the model, but it is often difficult to determine this constant. There are exceptions. Sometimes the DT is fixed at 1 as the phenomena being modeled occur on a periodic basis and data are limited to this time step. For example, certain insects may be born and counted on a given day each year. The DT is then 1 year and should not be reduced. The phenomenon is not continuous.

You may also change the numerical technique used to solve the model equations. Euler's method is chosen as a default. Two other methods, Runge–Kutta-2 and Runge–Kutta-4, are available to update state variables in different ways. These methods will be discussed later.

Start with a simple model and keep it simple, especially at first. Whenever possible, compare your results against measured values. Complicate your model only when your results do not predict the available experimental data with sufficient accuracy or when your model does not yet include all the features of the real system that you wish to capture. For example, we realize that the SICK do not remain so forever. Assume that they are sick for 25 days (SICK TIME) and then they get well. What is the new steady state level of SICK under these circumstances? To find the answer to this question, define an outflow from the stock SICK and name it GETTING WELL, the number of SICK who recover per day. There are at least two ways to evaluate this part of the model. We could just add the outflow GETTING WELL and then feed the SICK into it, and define GETTING WELL as 1/25 times the current level of SICK. This means that each day, 1/25th of the current level of SICK get well, and therefore, roughly, the average person would be sick for 25 days. Let us make this form of the addition first. Your model should look like this (Figure 1.14):

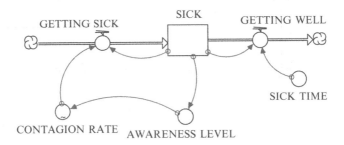

GETTING SICK SICK GETTING WELL SICK TIME CONTAGION RATE AWARENESS LEVEL

Fig. 1.14

Now the SICK disappear from the stock into a "cloud." You are not explicitly modeling where they go. Run the model again and find that the steady state number of SICK has dropped to 77 after about 90 days.

Now let us find a way to run a more accurate form of the model by converting our stock of SICK into a conveyor. A conveyor is really a symbol for a long series of connected stocks, in our case each with a one-day residence time. A conveyor must have the time spent in it, the Transit Time explicitly defined (here, 25 days). Use the dynamite symbol in the STELLA menu to remove the connector from SICK to GETTING WELL and the converter SICK TIME. When you dynamite a model component, put the fuse of the dynamite precisely over the object to be removed. For information arrow removal, put the fuse directly over the circle at the end of the arrow.

Open the stock SICK and click on conveyor, and then set Transit Time equal to 25. The initial value of the conveyor is set at 10, and STELLA will distribute this value evenly over its 25 time periods. If you want a more specific initial distribution, you must enter a value for each of the time periods, separated by a comma. For example, a conveyor with initial values of 2,1,3,.. would have a 2 as its initial value for the first time interval, a 1 for the second time interval, a 3 for the third, and so on.

Here is how your model should look now (Figure 1.15):

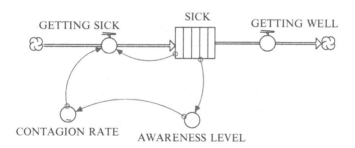

Fig. 1.15

Run the model and note the surprising result: damped cycling of the number of SICK. This is due to the explicit time delays introduced by the conveyor. Such delays are ever-present in the real world and introduce great complexity into the results of dynamic models. Here is a long-term picture of the results showing a long-term steady state of about 79 SICK, slightly higher than our previous, less accurate model (Figure 1.16):

The stock can also be turned into a queue or an oven. An oven allows entities to remain in a stock for a predetermined duration before they get released into a flow, and queues allows them to be queued up in their order of arrival. Each of these state variable forms is explained in the STELLA help file that comes with the full-implementation program.

You can of course continue to expand the model indefinitely, and the process is so easy and rewarding that you will be inclined to do so. But the goal of model is to

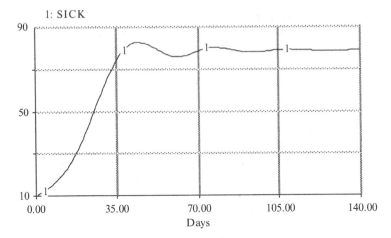

Fig. 1.16

create the simplest model that answers the questions you posed before you began! Otherwise, you will likely be adding unneeded complexity (and valuable effort and processing time) to your model.

As the model grows, it will contain an increasing number of submodels, or modules. You may want to protect some of these modules from being changed. To do this, click on the "Sector" symbol (to the left of the "A" in the next set of pictures) and drag it over the module or modules you want to protect. To run the individual sectors, go to "Sector Specs" in the RUN pull-down menu and select the ones that you wish to run. The values of the variables in the other sectors remain unaffected.

By annotating the model, you can remind yourself and inform others of the assumptions underlying your model and its submodels. This is important in any model, but especially in larger and more complicated models. To do this, click on the "Text" symbol (the letter A) and drag it into the diagram. Then type in your annotation. Notes can be left in the graph as well. Just click the "?".

The tools mentioned here are likely to prove useful when you develop more complicated models and when you want to share your models and their results with others. STELLA contains many helpful tools, which we hope you will use extensively. You will probably want to explore such features as "Drill Down" (visual hierarchy), "Space Compression," "High-Level Mapping Layer," "Arrays," and the "Authoring" features of STELLA. The appendix provides a brief overview of these and other features.

Make thorough use of your model, running it over again and always checking your expectations against its results. Change the initial conditions and try running the model to its extremes. At some point, you will want to perform a formal sensitivity analysis. The excellent sensitivity analysis procedures available in STELLA are discussed later.

1.6 Analogy and Creativity

If necessity is mother to invention, then analogy is father. Many new insights are generated by learning something from the structure or behavior of one entity, which is well understood, about another entity of which we have less knowledge. Leonardo da Vinci's observations of the forms and function of bones, tendons, and muscles of the human body, illustrated in his anatomical drawings, is analogous to the beam-bending device with supports and levers that he invented[3]. Newton's discovery of the law of gravity is thought to have been spurred by the realization that the apple and the moon are victims of the same forces, each within a different situation. Charles Babbage's invention of the "difference engine" that ultimately gave rise to the computer evoked images of a thinking machine[4]. Observations of the microscopic hierarchies of biological materials, ranging from the architecture of minerals and proteins in rats to the layers in chitin fibers in beetles are now used as biological analogues in the development of new, biomimicked materials or construction methods[5]. Recent attempts to generate artificial intelligence, using insight about the workings of the human brain, could be seen as still another extension of the human–machine analogy, this time in the realm of information processing rather than the purely material world[6].

Much of human reasoning is based on sparse data and the identification and comparison of patterns, instead of logical inference. We look at a new experience and try to match it with similar experiences in the past. A set of experiences and the patterns they form provide the basis for generalizations that then influence our decisions. Formation of analogies is a method to establish and organize correspondences among experiences and make them available for retrieval in the creative problem solving process.

There is often no strong logical base for mental models—at a deeper level, it is claimed, all thinking is metaphorical[7].

The interpolation and extrapolation among analogous patterns help form mental models that are often inadequate to provide a comprehensive perspective on the many interrelated aspects of systems and to anticipate their behavior—especially

---

[3] Mazlish, B., The Fourth Discontinuity: The Co-evolution of Humans and Machines, Yale University Press, New Haven, Connecticut, 1993.

[4] Penrose, R., The Emperor's New Mind: Concerning Computers, Minds, and the Laws of Physics, Penguin Books, 1989.

[5] Amato, I., Heeding the Call of the Wild, *Science*, Vol. 253, 1991, pp. 966–968.
Rubner, M. Synthetic Sea Shells, *Nature,* Vol. 423, 26 June 2003, pp. 925–926.

[6] Newby, T.J., P.A. Ertmer, and D.A. Stepich, *Instructional Analogies and the Learning of Concepts*, Educational Technology Research and Development, Vol. 48, No. 1, 1995, pp. 5–18.

[7] Lakoff, G. and M. Johnson, Metaphors We Live By, University of Chicago Press, Chicago, 1981, and:

——————, Philosophy in the Flesh: The Embodied Mind and Its Challenge to Western Thought, Basic Books, New York, 1999.

when we encounter novel situations[8]. While it is easy to show that mental models can be quite inaccurate, most of us do trust them in a wide range of (inappropriate) settings. This is why we must develop formal models. We must complement our thought processes and to reflect upon the workings and outcomes of formal models to sharpen our thinking. Formal models are based on an expressed logic, and they can form an analogic base—a repertoire of relationships on which to draw for problem solving.

Unfortunately, at least in their early years of education and employment, most people lack the analogic base of experience to draw on for their problem solving. This presents a kind of dilemma: if analogy is useful in problem solving and problem solving is fostered by drawing on analogies, how does one get the experience to build the analogic repertoire? The dilemma suggests an education plan where the analogic process itself is taught along with carefully structured problem solving. One may be exposed to classes of problems, each capable of being solved by generally the same analogic form. Later, the flaws of following the analogy too far can be pointed out and perhaps multiple analogies—each leading to a good solution form—can be demonstrated.

In any event, our book is structured to offer you a large set of modeling experiences, hoping that this set forms a sufficient experience base to launch you well into the world of modeling. With this set of modeling experiences, new problems can at first be broken into pieces that suggest analogies to items in this experience base. For example, our earlier models have analogies with water held temporarily in lakes, with the stock of capital in an industry, with a general population growth model, with a mechanical production line—the list seems endless. Once analogies are formed, the new problem begins to look solvable. As the model matures, the exact connection to the original analogies may be forgotten and a new modeling form may have been created.

1.7 STELLA's Numeric Solution Techniques

In this section we provide a brief description of, and basic mathematical background information on, the numeric techniques available in STELLA to solve the equations that define a model. Knowing about these techniques will be important in understanding how the computer arrives at model results, the accuracy of your results, and methods to improve accuracy.

From the models above we have seen that STELLA calculates the value of a stock in a given time period t based on the value of that stock a time step earlier plus the net of the inflows and outflows that occur over that time step. Generally, if $X(t)$ is the stock in time period t, $F(t, X(t), \cdot)$ are net flows that depend on time, the size of the stock $X(t)$ itself and possibly other parameters in the model (denoted by $\cdot$), and small time steps DT then

---

[8] Tversky, A. and D. Kahneman, The Framing of Decisions and the Psychology of Choice, *Science*, Vol. 211, 1981, pp. 453–458.

$$X(t) = X(t - DT) + F(t, X(t), \cdot) * DT \qquad (1.2)$$

Given an initial value $X(0)$, we can calculate $X(t)$ at any point in time as the sum of all the flows that occurred over all the small time steps DT between $t = 0$ and t:

$$X(t) = X(0) + \sum_{i=0}^{t} F(t, X(t), \cdot) * DT \qquad (1.3)$$

For example, in these models the number of SICK individuals today is a function of the population size one small time step earlier and the net additions that took place over that time step. This is the essence of equation (1.2). Equation (1.3) states that the number of SICK individuals after 140 days is the original population size plus all the net additions over the entire course of 140 days.

In these models, we chose DT = .25 and we specified in the Time Specs menu the "Numeric Method" to be Euler's method. Choosing Euler's method means that we used an equivalent to equation (1.2) to update the population size four times over the course of each day (see, for example, equation (1.3)).

Besides Euler's method, two other numeric solution techniques are available in STELLA. One of them is Runge-Kutta 2. With this method, stocks are updated in two steps as follows. First, a net flow F1 over the interval DT is calculated as with Euler's method:

$$F1 = F(t, X(t), \cdot) * DT \qquad (1.4)$$

Next, a second estimate F2 is generated by moving a small time step DT into the future:

$$F2 = F(t + DT, X(t) + F1, \cdot) * DT \qquad (1.5)$$

These two estimates are then used to calculate the stock $X(t)$ as

$$X(t) = X(t - DT) + 1/2 \, (F1 + F2) \qquad (1.6)$$

The second alternative to Euler's method that is available in STELLA for numeric approximation of flows and the updating of stocks is Runge-Kutta 4. Analogously to Runge-Kutta 2, Runge-Kutta 4 uses a set of four intermediate estimates to calculate $F(t, X(t), \cdot)$:

$$F1 = F(t, X(t), \cdot) * DT \qquad (1.7)$$
$$F2 = F(t + DT/2, X(t) + 1/2 F1, \cdot) * DT \qquad (1.8)$$
$$F3 = F(t + DT/2, X(t) + 1/2 F2, \cdot) * DT \qquad (1.9)$$
$$F4 = F(t, X(t) + F3, \cdot) * DT \qquad (1.10)$$

A weighted sum of those four estimates is then used to calculate the stock:

$$X(t) = X(t\text{-}DT) + 1/6 \, (F1 + 2 * F2 + 2 * F3 + F4) \qquad (1.11)$$

Numeric solution techniques such as the ones described above are often also called solution algorithms. How do these three algorithms compare with each other, and how does the choice of algorithm influence model results? Before we give an answer to this question, note that equation (1.6) can be used to express the net flow $F(t, X(t), \cdot)$ that occurs over a small time interval DT as the difference between the stock size at the beginning and end of a period of time. For example, equation (1.2) yields:

$$\frac{X(t) - X(t - DT)}{DT} = F(t, X(t), \cdot) \tag{1.12}$$

Equation (1.12) is known as a difference equation. It assumes that the stock X is updated over a discrete time interval DT.

Let us now define

$$\lim_{DT \to 0} = \frac{X(t) - X(t - DT)}{DT} \equiv \frac{dX}{dt} \tag{1.13}$$

then

$$\text{for } DT \to 0 \text{ we have } \frac{dX}{dt} = F(t, X(t), \cdot) \tag{1.14}$$

A calculation of dX/dt such as in (1.13) assumes an infinitesimally small time interval, and is known as a differential equation. The differential equation can be used to define the change in a stock X(t) as

$$dX = F(t, X(t), \cdot)dt \tag{1.15}$$

Analogously to equation (1.3), the stock X(t) in time t can be calculated for a given initial value of X(0) by summing up all the flows that occurred between time t = 0 and t:

$$x(t) = X(0) + \int_0^t F(u, X(u), \cdot)du \tag{1.16}$$

With these mathematical insights in mind, let us return to the comparison of the different numeric solution methods available in STELLA. If your model deals with changes in a system that is defined over continuous time, then a choice of DT significantly smaller than DT = 1 is required. Ideally, one would want DT to become infinitesimally small to do justice to the fact that time changes continuously, in infinitesimally small steps.

STELLA requires DT > 0 and will therefore solve all differential equations as difference equations. However, making DT very small gets us closer to a representation of changes in continuous time. Unfortunately, for a given numeric solution method and a given length of simulation the number of calculations needed to update the stocks increases as the size of DT is reduced.

If a system contains nonlinearities such as in Figure 1.17 and the DT is significantly larger than zero, approximation errors occur simply because the model keeps "jumping ahead in time" faster than is appropriate to keep track of the changes in system behavior that occur over the length of a DT. A smaller DT will minimize these errors but slow down the run of the model.

At a given DT, the Runge-Kutta 2 and Runge-Kutta 4 solution methods are typically more accurate than Euler's method because of the intermediate estimates made of F(t). Euler's method is fastest and Runge-Kutta 4 is slowest, however. This is because at a given choice of DT, more computational steps are needed with Runge-Kutta 4 then with Runge-Kutta 2, and more with Runge-Kutta 2 than with Euler's method.

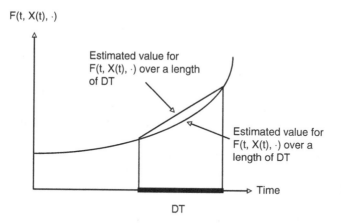

Fig. 1.17

Given the number of equations in each of the models of this book and given the use of state-of-the art computers, computation time is often not much of an issue. But if your computer is slow and your model is large, the choice of DT and solution method may have a significant impact on computation time. Your choice of DT and solution method will likely depend on the level of accuracy that you are willing or able to sacrifice for computation time. As we have mentioned before, a rule of thumb for this choice is to continue to decrease DT (or at a given DT switch to a more accurate solution method) until the changes in the results of your model fall within an acceptable limit. If for example, you reduce the DT of your model from .25 to .125 and the results of your model differ by less than a percent, while your input parameters are accurate only to about ± 5% and the model takes now twice as long to run, you may decide that cutting the DT was not worth the extra computational effort.

If your model is defined over discrete time, then you should choose a DT that is consistent with the length of the discrete time step. Experiment then with the choice of solution method to improve model accuracy.

Let us return to the disease model of Section 1.5 and investigate the sensitivity of the model results to the frequency at which we update the stock SICK in that model. Assume that the specification of this model presumes that the equations are defined over continuous time. The equations for GETTING SICK and GETTING WELL are thus differential equations (albeit they will be solved for DT > 0 and therefore be treated by STELLA as difference equations). For the model runs above, DT was set to .25. Generally speaking, a smaller DT leads to more accurate numerical calculation for updating state variables and, therefore, a more accurate answer. Also remember that if the flow equations must be understood as differential equations, DT needs to be sufficiently small.

Choose Time Specs from the RUN menu to change DT. Change DT to reflect ever-smaller periods until the change in the critical variable is within acceptable tolerances. Start with a DT = 1 and reduce it to 0.5, 0.25, 0.125 and so forth for subsequent runs, each time cutting it into half of its previous value. Before each run note the values of your state variable at the end of the previous run, as STELLA will erase the graphs if the DT is changed. To lock graphs, and thus not lose model results when you change the DT, click on the lock in the lower right-hand corner of the graph pad. This preserves the results. Then double-click on the graph pad to open its dialog box, click on the upward-pointing triangle to generate a new page for the graph pad, and select the variable(s) to be plotted. Now, one page of the graph pad contains the results for one choice of DT; the other contains the results after changing DT.

To get an even more accurate reading of your model results at each point in time, create a table. Choose the table icon in STELLA, place it in the model diagram, double-click anywhere on the table and select SICK as the input, then double-click on the head of the column where it reads "SICK" and choose "Free Float" as the "Precision." With this specification, your table will report the results with an accuracy of more than the two decimal places that would otherwise be listed. (In this case, though, reporting the number of individuals who are sick with more accuracy than whole numbers makes little sense, unless the unit of measurement is, for example, in millions of people.) However, irrespective of the level of precision at which you report the results, the computation itself is not affected by that choice. Click OK and run the model. Note the results (perhaps lock the results of this page of the table, much as you would lock results on a page of a graph), and then proceed to change the DT to a smaller value. Compare the results from one model run to the next. Repeat this process for different solution methods, and observe changes in model errors. Other sources of errors are discussed in more detail in the following section.

1.8 Sources of Model Error

Error is associated with virtually every aspect of a model. As we have discussed above there are errors involved in the algorithms—and sometimes even at the hardware level of the computer—used to numerically solve the model. A set of errors is also associated with the conceptualization of the model and is the topic of this section[9].

Any model is an abstraction. In the process of making this abstraction, a limited set of system components and their interactions are considered. Features that are irrelevant to the system's dynamics over the temporal and spatial range or to the

---

[9] The discussion of sources of errors presented in this section follows Westervelt (2001).

questions that the model addresses are disregarded. Errors of exclusion come from overlooking a particular system feature that *does* have a relevant influence on the dynamics of the system components that are explicitly modeled, and thus has a relevant influence on the results of the model. A model of the impacts of a disease on the dynamics of an animal population may assume constancy of the environment within which the animals live. However, the population itself may alter its environment, such as through removal of prey species. As a consequence, an increase in morbidity or mortality in the population on which our model concentrates may trigger fundamental changes in the ecosystem and possibly make the various parameters on which the model is based inadequate to capture what is really going on.

Errors of inclusion are associated with explicitly modeling aspects of the system that are irrelevant for understanding its dynamics, and have no bearing on model result. Unnecessary effort goes into those parts without corresponding gain. At a given modeling budget or time frame available for the completion of the model, including unnecessary parts may mean that the important parts are not modeled to their fullest detail or extent, and that those essential parts are therefore more prone to inaccuracies or errors.

Models are often based on scientific facts that have been established on the basis of controlled experiments, as we have discussed in the previous chapter. Models may also include less formal knowledge of stakeholders, derived on the basis of their personal experience, anecdotal information, or collective heritage. In either case, the resolution or temporal scale to which the formal or informal knowledge applies may be inconsistent with the resolution and scale of the computer model. Errors of interpolation and extrapolation may exist by applying formal or informal knowledge outside the domains for which the knowledge has been established.

There are two main errors of inappropriate temporal specification. One is caused from not running the model long enough. When the model contains nonlinearities and time lags, some of the dynamics may not unfold within a short time frame. Running the model over an extended temporal range can easily reduce errors of inappropriately truncating model dynamics.

The other main error associated with the temporal specification of the model is related to the choice of DT. Frequently, scientific studies of marine systems make use of differential equations and assume that DT is infinitesimally small. The choice of differential equations is driven by the desire to solve for key system properties, such as their steady state conditions. Those solutions are derived analytically by applying calculus of variations. Using differential equations from scientific studies within a computer model that numerically solves for the system states at each period of time means that $DT > 0$. This leads to inconsistencies with the original studies on which the model is based, and to approximation errors as discussed above. Choosing very small DT for modeling differential equations and exploring model sensitivity to the choice of DT can help reduce errors of inappropriate choice of DT.

Similar to errors of inappropriate temporal specifications, there are two main errors of inappropriate spatial specification—spatial boundary effects and inappropriate spatial resolution. Boundary effects are related to errors of exclusion and are caused by assuming that what lies outside the boundaries drawn around the modeled area does not influence the dynamics within the area. This assumption is obviously quite critical if we wish to model, for example, a nature preserve that is temporarily visited by highly mobile species. The value that the preserve provides for those species may be rather limited if residence times within the preserve are small and human-induced mortality rates outside the preserve are very high. To reduce errors from spatial boundary assumptions, spatial modelers often draw spatial boundaries a bit larger than they know they actually need or set up special rules for the processes along the boundary (to mimic the interactions between the system that is explicitly modeled with what lies outside that system).

Errors of inappropriate spatial and temporal resolution can be related to each other. For example, a model used to trace the movement of individuals in a population may subdivide the area into adjacent grid cells and then specify the decision rules by which movement from one cell to the next occurs. If the population is highly mobile, movement in a given time step can be further than the resolution of the space, and errors of spatial resolution result.

The computer model requires that initial conditions and parameter values be specified for a point in time. Those initial conditions and parameter values are often derived on the basis of field or laboratory measurements, and they are typically fraught with errors. Good empirical work should report confidence intervals for the measurements, and a good model should explore a system's dynamics at least within the reported range of confidence intervals to minimize errors of model inputs.

Once the model is specified, the difference equations are solved by the computer in a specific order. To see the order for execution of equations in your STELLA model, navigate with the downward-pointing triangle to the model's equation window, then choose "Equation Prefs" from the Equation pull-down menu and select "Order of Execution." There is nothing you can do to influence this order within STELLA once all your equations are defined, but note that the choice of order may introduce errors of aninappropriate order of execution. For example, if in a spatial model of migrating individuals the death rate is a function of population density and density is computed before migration occurs, then death rates (deaths per time unit) will be different from the case in which density is calculated after migration. Keep this in mind when you specify your model, and if necessary introduce time lags to achieve the desired order in calculations.

Your modeling effort should start with a clear question in mind. The choices of system components that you wish to model, spatial and temporal resolution, data sources, solution method, and DT should be driven by that question. Avoid having these choices be driven by the answer that you expect and wish to generate.

At some point of your modeling career, you may find that you are so excited by your model results that you overextend the conclusions; for example, by describing the dynamics you see with words such as *never* or *always*. Even if you have done all you can to base your model on the best available knowledge, earlier discussion of the various sources of errors should highlight the danger of making errors by drawing inappropriate conclusions.

The following section provides a set of guidelines designed to facilitate model development and help you avoid errors. An important recommendation is to explore the sensitivity of model results to different model specifications. Later chapters will take up the issue of sensitivity analysis in more detail.

If enough data are available, check your model's ability to use known initial data and "predict" known recent data for the same variables. This calibration and verification step lets you project scenarios with greater certainty.

1.9 The Detailed Modeling Process

Following is a set of easy-to-follow but detailed modeling steps. These steps are not sacred: they are intended as a guide to get you started in the process. You will find it useful to come back to this list once in a while as you proceed in your modeling efforts.

1. Define the problem and the goals of the model. Frame the questions you want to answer with the model. If the problem is a large one, define subsystems of it and goals for the modeling of these subsystems. Ask yourself: Is my model intended to be explanatory or predictive?
2. Designate the state variables. (These variables will indicate the status of the system.) Keep it simple. Purposely avoid complexity in the beginning. Note the units of the state variables.
3. Select the control variables—the flow controls into and out of the state variables. (The control variables are calculated from the state variable in order to update them at the end of each time step.) Note to yourself which state variables are donors and which are recipients with regard to each of the control variables. Also, note the units of the control variables. Keep it simple at the start. Try to capture only the essential features. Put in one type of control as a representative of a class of similar controls. Add the others in step 10.
4. Select the parameters for the control variables. Note the units of these parameters and control variables. Ask yourself: Of what are these controls and their parameters a function?
5. Examine the resulting model for possible violations of physical, economic, and other laws (for example, any continuity requirements or the conservation of mass, energy, and momentum). Also, check for consistency of units. Look for the possibilities of division by zero, negative volumes or prices, and so forth. Use conditional statements if necessary to avoid these violations.
6. To see how the model is going to work, choose some time horizon over which you intend to examine the dynamic behavior of the model, the length of each time interval for which state variables are being updated, and the numerical computation procedure by which flows are calculated. (For example, choose in the STELLA program Time Step = 1, time length = 24.) Set up a graph and guess the variation of the state variable curves before running the model.
7. Run the model. See if the graph of these variables passes a "sanity test." Are the results reasonable? Do they correspond to known data? Choose alternative

lengths of each time interval for which state variables are updated. Choose alternative integration techniques. (In the STELLA program, for example, reduce the time interval DT by half and simulate the mode again to see if the results are the same.)

8. Vary the parameters to their reasonable extremes and see if the results in the graph still make sense. Revise the model to repair errors and anomalies.

9. Compare the results to experimental data. This may mean shutting off parts of your model to mimic a lab experiment, for example.

10. Revise the parameters, perhaps even the model, to reflect greater complexity and to meet exceptions to the experimental results, repeating steps 1–10. Frame an enlarged set of further questions. Consider the analogies to your model. Can these analogies further inform your model?

Do not worry about applying all of these steps in this order as you develop your models and improve your modeling skills. Do check back to this list now and then to see how useful, inclusive, and reasonable these steps are.

You will find that modeling has three possible general uses. First, you can experiment with models. A good model of a system enables you to change its components and see how these changes affect the rest of the system. This insight helps you explain the workings of the system you are modeling. Second, a good model enables prediction of the future course of a dynamic system. Third, a good model stimulates further questions about the system behavior and the applicability of the principles that are discovered in the modeling process to other systems.

Remember the words of Walter Deming: "All models are wrong. Some are useful." To this we add: "No model is ever complete."

1.10 Questions and Tasks

1. We have stated that the key attributes of a model aspiring to reality are feedbacks, delays, and randomness. The earlier models show the feedbacks and delays. Can you add uniform randomness to the multiplier in AWARENESS LEVEL?

2. Add an explicit delay in the GETTING SICK control variable in Figure 1.14. Try different levels of delay to see if you can get a permanent cycle in the SICK population.

3. Suppose that AWARENESS was a function in part of the SICK TIME. How would you represent such a possibility in the model in Figure 1.15?

Chapter 2
Basic Epidemic Models

2.1 Basic Model

In this chapter we follow up on our discussion from Chapter 1 and model the spread of a disease through a population, gradually adding new features. Epidemics, such as the one modeled here, are of great concern to human societies. The complex interrelationships of biological, social, economic and geographic relationships that drive or constrain an epidemic make dynamic models an invaluable tool for the analysis of particular diseases. The model developed here is fairly idealized but can be applied easily to real populations affected by a disease[1].

Assume that an initial population of 1,000,000 (per 100 square miles) is not immune to a contagious disease. The rate at which they become sick is assumed to be a function of the product of the nonimmune population times the contagious plus sick population. This equation for contagion is the simplest form that meets the obvious requirements that the contagion rate must be zero if either the immune or the contagious populations are zero. The contagious population is assumed to become the sick population for a week and then, with a survival rate of 0.9, the survivors join the immune population. The nonimmune population is augmented with a constant birth rate of 5000/week. The people in this model do not die of other causes.

Setting the contagion rate proportional to the product of the nonimmune and the contagious and sick population is arbitrary. The form is suspiciously similar

---

[1] See Spain, J.D. 1982. *BASIC Microcomputer Models in Biology*, Addison-Wesley, Reading, Massachusetts, p. 118. For some realism, see the data on the Black Death in 1300s Italy (Curtis H. and N. Barnes. 1985. *Invitation to Biology*, Worth Publishers, New York.) These data show a declining peak as people became aware of the vector, or those most likely exposed to the vector died off, or the naturally immune were selected for and that immunity was inheritable. The four occurrences of the plague in that century had a period of about eleven years. For chaotic epidemics, see: Schaffer, W. 1985. Can Nonlinear Dynamics Elucidate Mechanisms in Ecology and Epidemiology, *IMA Journal of Mathematics Applied in Medicine and Biology*, Vol. 2, pp. 221–252. Schaffer shows how a cyclic contact coefficient can produce chaos in this form of epidemic model.

B. Hannon and M. Ruth, *Dynamic Modeling of Diseases and Pests*,
Modeling Dynamic Systems,
© Springer Science + Business Media LLC 2009

to the way chemical reactions can be specified with the law of mass action[2]—the concentration of a chemical product (e.g. number of moles of a substance per cubic meter air) is computed as the product of the concentrations of the reactants and a reaction rate. Here, we convert the currently nonimmune population into a sick population:

SICK RATE = CONTACT RATE * (CONTAGIOUS+SICK) * NONIMMUNE.

$$(2.1)$$

The structure of the complete model is shown in Figure 2.1. Initialize this model so you have a starting populations of NON IMMUNE = 1000000, CONTAGIOUS = 1, SICK = 0, and IMMUNE = 0. Assume a fixed number of births of 5000 per week and a contact rate of .000002. Also, assume that individuals who contract the disease will on average be moving around for one week before they are bedridden and that they are contagious during that time as well as the time during which they are confined to bed. The same contact rate applies to both subsets of the population. Furthermore, as noted above, assume that every week 90 percent of those sick individuals, who are confined to their beds, will recover and become permanently immune; the remainder of them die.

The outcome for such a simple model (Figure 2.2) is interesting. It would be difficult to predict that the simple equation for GET SICK would yield the remarkably authentic series of diminishing pulses. In many ways, the appearance of these pulses is an emergent property of the model, and it is realistic. The initial epidemic is the most severe and converts 90% of the population to an immune condition.

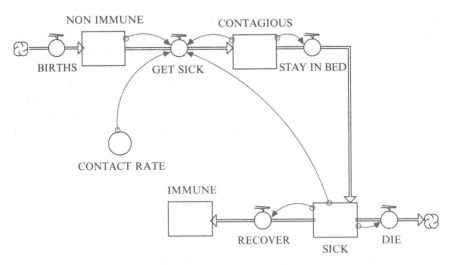

Fig. 2.1

---

[2] See, for example, Hannon, B. and M. Ruth. 2001. Dynamic Modeling, 2nd Edition, Springer Verlag, New York.

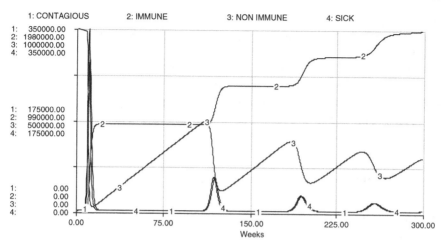

Fig. 2.2

Ensuing epidemics occur with regular frequency and are increasingly less severe, finally reaching a steady-sized nonimmune population. (Can you explain why this is so?) The disease has become endemic. At this steady state, the immune population is growing at the birth rate, and the contagion rate is constant and equal to the birth rate.

BASIC MODEL

CONTAGIOUS(t) = CONTAGIOUS(t − dt) + (GET_SICK − STAY_IN_BED) * dt
INIT CONTAGIOUS = 1 {Individuals}

INFLOWS:
GET_SICK = CONTACT_RATE * (CONTAGIOUS + SICK) * NON_IMMUNE
{Individuals per Time Period}
OUTFLOWS:
STAY_IN_BED = CONTAGIOUS {Individuals per Time Period}
IMMUNE(t) = IMMUNE(t − dt) + (RECOVER) * dt
INIT IMMUNE = 0 {Individuals}

INFLOWS:
RECOVER = .9 * SICK {Individuals per Time Period}
NON_IMMUNE(t) = NON_IMMUNE(t − dt) + (BIRTHS − GET_SICK) * dt
INIT NON_IMMUNE = 1000000 {Individuals}

INFLOWS:
BIRTHS = 5000 {Individuals per Time Period}
OUTFLOWS:
GET_SICK = CONTACT_RATE * (CONTAGIOUS + SICK) * NON_IMMUNE
{Individuals per Time Period}
SICK(t) = SICK(t − dt) + (STAY_IN_BED − RECOVER − DIE) * dt
INIT SICK = 0 {Individuals}

INFLOWS:
STAY_IN_BED = CONTAGIOUS {Individuals per Time Period}
OUTFLOWS:
RECOVER = .9 * SICK {Individuals per Time Period}
DIE = .1 * SICK {Individuals per Time Period}
CONTACT_RATE = .000002 {1/(Number of Contagious + Sick) *
Nonimmune) per Time Period}

2.2 Epidemic Model with Randomness

Let us assume for the model of the previous section that the contact rate at any given week may be somewhere between .000001 and .000003. Thus, instead of fixing the contact rate to .000002, as we have done before, we now let it vary randomly between its two extremes. The built-in function RANDOM can be used to specify the contact rate accordingly. Open CONTACT RATE and choose RANDOM from the list of built-in functions, or type it into the dialog box. Then specify the upper and lower bounds, separating the entries by a comma. The model should now read:

$$\text{CONTACT RATE} = \text{RANDOM}(.000001, .000003) \qquad (2.2)$$

The rest of the model should stay the same, as in Figure 2.1. Will this change in model specification significantly affect the periodicity or severity of the epidemic outbreaks? See the results in Figure 2.3 and note that the behavior of the disease is virtually the same for the first outbreak of the disease, but that the randomness of the contact rate has a much more noticeable impact for smaller outbreaks, which on

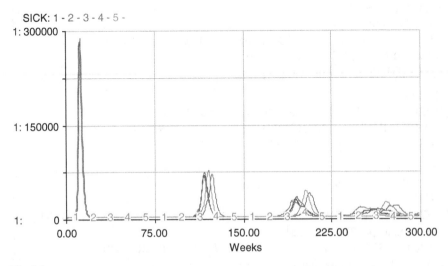

Fig. 2.3

some occasions occur later, on others earlier. Also, their severity varies. Can you explain why this is so?

The results in Figure 2.3 are for five model runs. We have checked the "Comparative" box in the graph pad to enable multiple runs to be displayed in the same page of the graph pad. Do you expect the model results for a large number of runs—on average—yield the same pattern as we observed in the previous section?

If you wish to let a parameter vary from run to run along a normal distribution with known mean and standard deviation, choose the built-in function NORMAL and specify a mean and standard deviation. Other specifications of random variables are available as well. Make use of them as you see fit.

Both the RANDOM and NORMAL built-ins allow for the option to specify a "seed," which ensures that STELLA uses for subsequent runs the same sequence of random numbers as it does the first time the model is run with a seed. For example, if you have two parameters that are specified as random numbers, say A and B, and you specify

$$A = RANDOM(0, 1, 1) \qquad (2.3)$$
$$B = RANDOM(0, 1, 2) \qquad (2.4)$$

then subsequent runs of the model use the "first" strings of random numbers (all between zero and one) generated for A and the "second" string of random numbers (also between zero and one) generated for B. If you do not specify a seed, then all subsequent runs will differ simply because different random numbers will be chosen. While this may be informative in some cases, as for Figure 2.3, it is sometimes difficult to see whether changes in a model's results stem the mere fact that the model uses different random numbers or whether they stem from the fact that the model itself was changed.

EPIDEMIC WITH RANDOMNESS

CONTAGIOUS(t) = CONTAGIOUS(t − dt) + (GET_SICK − STAY_IN_BED) * dt
INIT CONTAGIOUS = 1 {Individuals}

INFLOWS:
GET_SICK = CONTACT_RATE * (CONTAGIOUS + SICK) * NON_IMMUNE {Individuals per Time Period}
OUTFLOWS:
STAY_IN_BED = CONTAGIOUS {Individuals per Time Period}
IMMUNE(t) = IMMUNE(t − dt) + (RECOVER) * dt
INIT IMMUNE = 0 {Individuals}

INFLOWS:
RECOVER = .9 * SICK {Individuals per Time Period}
NON_IMMUNE(t) = NON_IMMUNE(t − dt) + (BIRTHS − GET_SICK) * dt
INIT NON_IMMUNE = 1000000 {Individuals}

INFLOWS:
BIRTHS = 5000 {Individuals per Time Period}

OUTFLOWS:
GET_SICK = CONTACT_RATE * (CONTAGIOUS + SICK) * NON_IMMUNE
{Individuals per Time Period}
SICK(t) = SICK(t − dt) + (STAY_IN_BED − RECOVER − DIE) * dt
INIT SICK = 0 {Individuals}

INFLOWS:
STAY_IN_BED = CONTAGIOUS {Individuals per Time Period}
OUTFLOWS:
RECOVER = .9 * SICK {Individuals per Time Period}
DIE = .1 * SICK {Individuals per Time Period}
CONTACT_RATE = RANDOM(.000001,.000003) {1/(Number of Contagious +
Sick) * Nonimmune) per Time Period}

Recall that the three major aspects of any model for which reality is claimed
should have feedback, randomness, and delays. Our model now has all of these aspects: feedback from the CONTAGIOUS to GETTING SICK, randomness in the
CONTACT RATE, and delays in STAYING IN BED and RECOVERING. The use
of stocks for CONTAGIOUS and SICK provide automatic one-week delays in each
of these stocks since the outflows are divided by one; to increase this type of delay, we could divide these outflows by larger numbers. We could also change these
stocks into conveyors as we did in Chapter 1.

2.3 Loss of Immunity

How will the dynamics of our epidemic change if people lose immunity? Let us
assume for the model of Section 2.1 above that in any given week, 10 percent of the
immune population loses their immunity. All other assumptions are as before. The
corresponding model is shown in Figure 2.4.

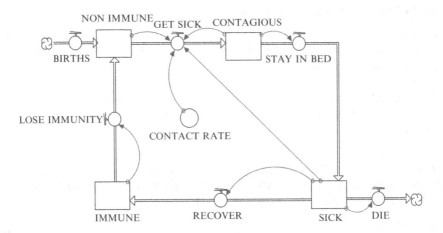

Fig. 2.4

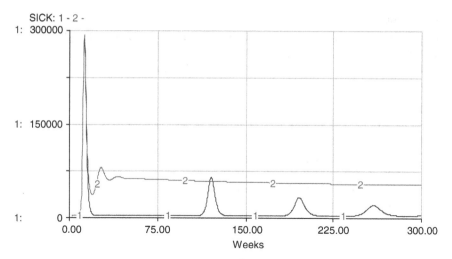

Fig. 2.5

The model results are shown in Figure 2.5. The loss of immunity dampens the effect of the disease—outbreaks after the first occurrence are less severe and in the long run, the number of sick people is constant. That number is typically larger than would be the case of permanent immunity.

LOSS OF IMMUNITY

CONTAGIOUS(t) = CONTAGIOUS(t − dt) + (GET_SICK − STAY_IN_BED) * dt
INIT CONTAGIOUS = 1 {Individuals}

INFLOWS:
GET_SICK = CONTACT_RATE * (CONTAGIOUS + SICK) * NON_IMMUNE
{Individuals per Time Period}
OUTFLOWS:
STAY_IN_BED = CONTAGIOUS {Individuals per Time Period}
IMMUNE(t) = IMMUNE(t − dt) + (RECOVER − LOSE_IMMUNITY) * dt
INIT IMMUNE = 0 {Individuals}

INFLOWS:
RECOVER = .9*SICK {Individuals per Time Period}
OUTFLOWS:
LOSE_IMMUNITY = .1*IMMUNE
NON_IMMUNE(t) = NON_IMMUNE(t − dt) + (BIRTHS +
LOSE_IMMUNITY − GET_SICK) * dt
INIT NON_IMMUNE = 1000000 {Individuals}

INFLOWS:
BIRTHS = 5000 {Individuals per Time Period}
LOSE_IMMUNITY = .1*IMMUNE

OUTFLOWS:
GET_SICK = CONTACT_RATE * (CONTAGIOUS + SICK)*NON_IMMUNE
{Individuals per Time Period}
SICK(t) = SICK(t − dt) + (STAY_IN_BED − RECOVER − DIE) * dt
INIT SICK = 0 {Individuals}

INFLOWS:
STAY_IN_BED = CONTAGIOUS {Individuals per Time Period}
OUTFLOWS:
RECOVER = .9 * SICK {Individuals per Time Period}
DIE = .1 * SICK {Individuals per Time Period}
CONTACT_RATE = .000002 {1/(Number of Contagious + Sick) *
Nonimmune) per Time Period}

2.4 Two Population Epidemic Model

Many diseases can affect different species and be spread from one species to another.
Notable examples include AIDS and Ebola. To concurrently capture the spread of
a disease within and between species, we will use a simplified version of our basic
model of Section 2.1 above. Let's first lump the CONTAGIOUS and SICK populations into one stock, and call it INFECTED and then define a SURVIVAL RATE,
which we set to 0.065. The flows SURVIVE and DIE are specified as

$$SURVIVE = SURVIVAL\_RATE * INFECTED \qquad (2.5)$$

$$DIE = (1 − SURVIVAL\_RATE) * INFECTED \qquad (2.6)$$

This is a rather lethal disease compared to the one in Section 2.1. The new model
is shown in Figure 2.6 and its dynamics are shown in Figure 2.7.

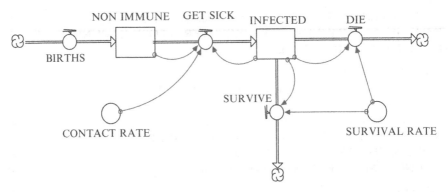

Fig. 2.6

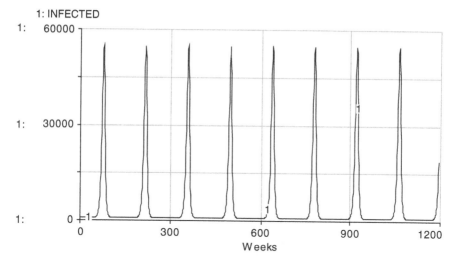

Fig. 2.7

To capture the spread of disease in the second population, duplicate the model of Figure 2.6 by first clicking outside its boundaries and dragging the Hand over the entire model diagram. Then select Copy from the Edit pull-down menu. Then click well below the existing model diagram and select Paste from the Edit pull-down menu. Here is what you should get—two virtually identical versions of the model, with the names of the second model automatically changed to avoid confusion (Figure 2.8). Change the value of SURVIVAL RATE 2 to 0.20 and that of CONTACT RATE 2 to 0.003 to reflect the assumptions that the rate of contact among individuals of this second species are higher but the death rate lower than for the first species. This is the case, for example for some strains of the AIDS and Ebola viruses, which seem, at least until these viruses evolve, to have more detrimental effects on humans than monkeys.

The two parts of the model in Figure 2.8 are not yet connected with each other. Create a new contact rate to reflect interaction between the two populations, and assume that the disease gets passed on from the second to the first population, i.e.

$$\text{GET SICK} = \text{CONTACT\_RATE} * \text{NON\_IMMUNE} * \text{INFECTED}$$
$$+ \text{CONTACT\_RATE\_1\_2} * \text{NON\_IMMUNE} * \text{INFECTED\_2}$$
$$(2.7)$$

Use this contact rate to establish a logical connection between the two parts of the model. Figure 2.9 shows incorporation of that new contact rate and the corresponding connections between GET SICK and the nonimmune population of the second species.

If you set CONTACT RATE 1 2 to 0, then the dynamics of the disease in the first population should be as before (i.e. as in Figure 2.7). For subsequent runs explore how the results change under different assumptions about the rate of contact

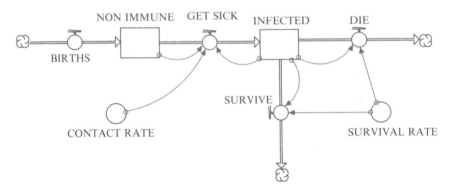

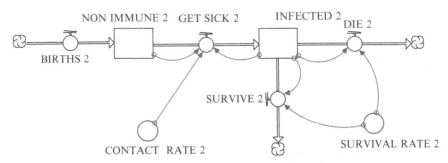

Fig. 2.8

between the two populations. You can "automate" this process by choosing Sensi Specs under the Run pull-down menu. Then select CONTACT RATE 1 2 to learn how sensitive the model results are to different parameter values. Select Incremental and specify the start value for your sensitivity analysis as .001 and your end value as 0.0014. If the number of sensitivity runs is set to 3, then STELLA will run the model first with the start value, second with a value that lies equidistant between your start and end value (i.e. 0.0012 in our case), and last with the end value you specified. Click OK. Then run the model. Note how STELLA now shows S-Run in the Run pull-down menu to indicate that this will be a sensitivity run. The results are shown in Figure 2.10. The first two runs yield period outbreaks of the disease, where the second run shows less frequent but more severe outbreaks. Also, while the severity of the outbreaks for the first run slightly declines in the long term, it increases for the second run. In contrast to these results, the disease disappears for the third model run. The contact rate for interactions between the two populations is so high that the disease "burns out" after the first outbreak.

If you wish for a sensitivity analysis to let parameters vary from run to run along a normal distribution with known mean and standard deviation, choose the "Distribution" option (Figure 2.11) instead of "Incremental." When you specify "Seed" as a positive integer, you ensure the ability to replicate a particular random number sequence in subsequent sensitivity runs, just like we have done in

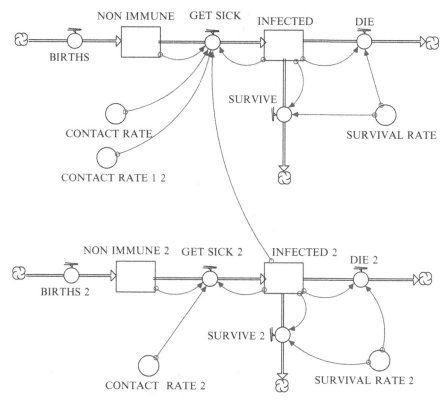

Fig. 2.9

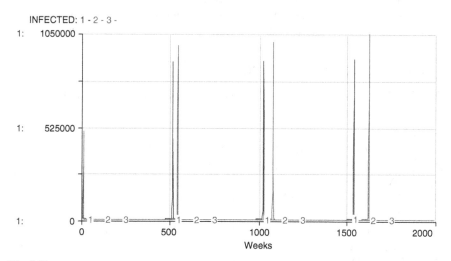

Fig. 2.10

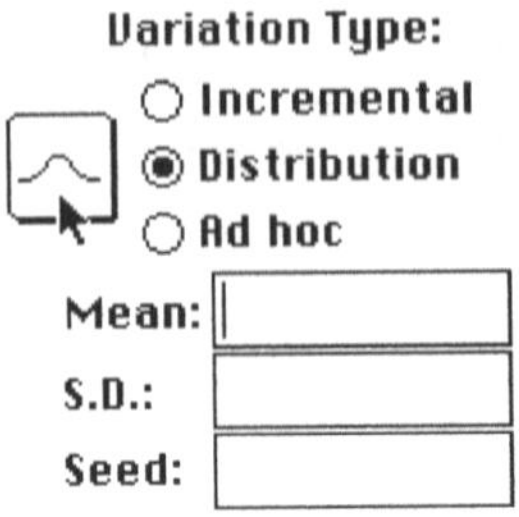

Uariation Type:

○ Incremental
● Distribution
○ Ad hoc

Mean: |

S.D.:

Seed:

Fig. 2.11

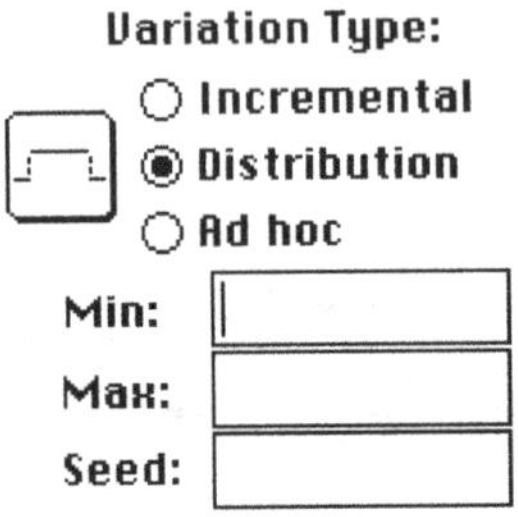

Uariation Type:

○ Incremental
● Distribution
○ Ad hoc

Min: |

Max:

Seed:

Fig. 2.12

Section 2.2. If you do not wish to make use of the normal distribution of the random numbers used in the sensitivity analysis, click on the bell curve button.

This button will change its appearance (Figure 2.12), and you now need to specify a minimum, maximum, and seed for your sensitivity analysis.

A final choice for the specification of sensitivity runs in STELLA is not to change parameters in incremental intervals or along distributions. You can specify *ad hoc* values for each of the consecutive runs.

TWO POPULATION EPIDEMIC MODEL

INFECTED(t) = INFECTED(t − dt) + (GET_SICK − SURVIVE − DIE) * dt
INIT INFECTED = 1

INFLOWS:
GET_SICK = CONTACT_RATE * NON_IMMUNE * INFECTED +
CONTACT_RATE_1_2 * NON_IMMUNE * INFECTED_2
OUTFLOWS:
SURVIVE = SURVIVAL_RATE * INFECTED
DIE = (1 − SURVIVAL_RATE) * INFECTED
INFECTED_2(t) = INFECTED_2(t − dt) + (GET_SICK_2 − DIE_2 −
SURVIVE_2) * dt
INIT INFECTED_2 = 20

INFLOWS:
GET_SICK_2 = CONTACT__RATE_2 * INFECTED_2 * NON_IMMUNE_2

OUTFLOWS:
DIE_2 = (1 − SURVIVAL_RATE_2) * INFECTED_2
SURVIVE_2 = SURVIVAL_RATE_2 * INFECTED_2
NON_IMMUNE(t) = NON_IMMUNE(t − dt) + (BIRTHS − GET_SICK) * dt
INIT NON_IMMUNE = 1000000

INFLOWS:
BIRTHS = 5000
OUTFLOWS:
GET_SICK = CONTACT_RATE * NON_IMMUNE * INFECTED +
CONTACT_RATE_1_2 * NON_IMMUNE * INFECTED_2
NON_IMMUNE_2(t) = NON_IMMUNE_2(t − dt) + (BIRTHS_2 −
GET_SICK_2) * dt
INIT NON_IMMUNE_2 = 1000

INFLOWS:
BIRTHS_2 = 10
OUTFLOWS:
GET_SICK_2 = CONTACT_RATE_2 * INFECTED_2 * NON_IMMUNE_2
CONTACT_RATE = .000001
CONTACT_RATE_1_2 = 0.00015
CONTACT_RATE_2 = .003
SURVIVAL_RATE = .065
SURVIVAL_RATE_2 = .2

2.5 Epidemic with Vaccination

The model of this section introduces a number of features that make the model more meaningful. Among these features are

- the explicit inclusion of birth rates (instead of applying a fixed number of births each period);
- death rates that are not only influenced by the disease but that result also from natural mortality;
- a vaccination program that allows the population to become immune to the disease without having to first be sick;
- mutations in the disease that result in immune people not staying immune forever; and
- ignorance of a fixed portion of the contagious population. These people are assumed to be unaware that they carry the disease. Consequently, we assume that ignorance would increase the rate at which the disease gets passed on from the infective to the susceptible population.

The birth and natural death rates are specified graphically in this model as shown in Figures 2.13 and 2.14.

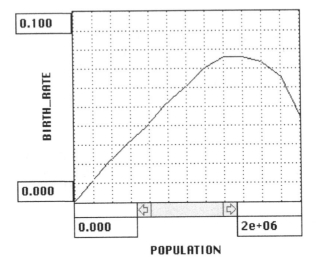

Fig. 2.13

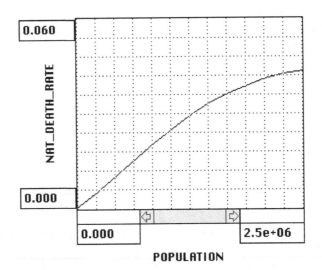

Fig. 2.14

The CONTRACTION is a function of susceptible, healthy people coming into contact with people who are aware that they are contagious or people that are unaware that they are contagious, or people who come into contact with sick people. BETA is the corresponding contact rate:

$$\text{CONTRACTION} = \text{BETA}^*((\text{CONTAGIOUS-UNAWARE CONTAGIOUS})^*3/5$$
$$+ \text{UNAWARE CONTAGIOUS+SICK}/2)^*\text{NONIMMUNE}$$
$$(2.8)$$

The VACCINE flow is specified as

$$\text{VACCINE} = \text{NON IMMUNE}^*(1 - \text{CONTRACTION}) \qquad (2.9)$$

and contains the number of people vaccinated. A flow from IMMUNE to NON-IMMUNE, captures 1.5 percent of immune people that lose their immunity (Figure 2.15).

Consistent with the models of Sections 2.1 and 2.3, there are a series of epidemic outbreaks (Figure 2.16). Due to the additional features of this model, however, the numbers of immune and nonimmune people tends to stabilize and so does the number of sick. Even though an effective immunization program is in place,

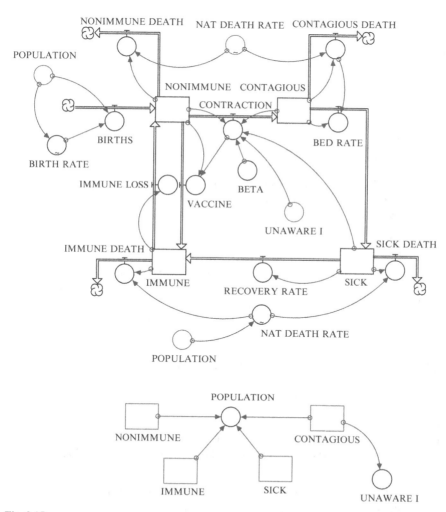

Fig. 2.15

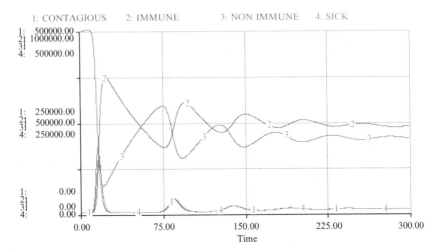

Fig. 2.16

some sick people are always present because the disease is assumed to undergo mutations.

EPIDEMIC WITH VACCINATION

CONTAGIOUS(t) = CONTAGIOUS(t − dt) + (CONTRACTION − BED_RATE − CONTAGIOUS_DEATH) * dt
INIT CONTAGIOUS = 1 {People}

INFLOWS:
CONTRACTION =
BETA * ((CONTAGIOUS − UNAWARE_CONTAGIOUS) * 3/5 + UNAWARE_CONTAGIOUS + SICK/2) * NONIMMUNE {People per Week}
OUTFLOWS:
BED_RATE = CONTAGIOUS − CONTAGIOUS_DEATH {People per Week}
CONTAGIOUS_DEATH = CONTAGIOUS * NAT_DEATH_RATE/52 {People per Week}
IMMUNE(t) = IMMUNE(t − dt) + (RECOVERY_RATE + VACCINE − IMMUNE_DEATH − IMMUNE_LOSS) * dt
INIT IMMUNE = 0 {People}

INFLOWS:
RECOVERY_RATE = 0.9 * SICK {People per Week}
VACCINE = NONIMMUNE * (1 − CONTRACTION) {People per Week}

OUTFLOWS:
IMMUNE_DEATH = IMMUNE * NAT_DEATH_RATE/52 {People per Week}
IMMUNE_LOSS = .015 * IMMUNE {People per Week}

NONIMMUNE(t) = NONIMMUNE(t − dt) + (BIRTHS + IMMUNE_LOSS − CONTRACTION − NONIMMUNE_DEATH − VACCINE) * dt
INIT NONIMMUNE = 1000000 {People}

INFLOWS:
BIRTHS = POPULATION * BIRTH_RATE/52 {People per Week}
IMMUNE_LOSS = .015 * IMMUNE {People per Week}
OUTFLOWS:
CONTRACTION =
BETA * ((CONTAGIOUS − UNAWARE_CONTAGIOUS) * 3/5 + UNAWARE_CONTAGIOUS + SICK/2) * NONIMMUNE {People per Week}
NONIMMUNE_DEATH = NAT_DEATH_RATE/52 * NONIMMUNE {People per Week}
VACCINE = NONIMMUNE * (1 − CONTRACTION) {People per Week}
SICK(t) = SICK(t − dt) + (BED_RATE − RECOVERY_RATE − SICK_DEATH) * dt
INIT SICK = 0 {People}

INFLOWS:
BED_RATE = CONTAGIOUS-CONTAGIOUS_DEATH {People per Week}
OUTFLOWS:
RECOVERY_RATE = 0.9 * SICK {People per Week}
SICK_DEATH = (.1 * SICK) + (NAT_DEATH_RATE/52 * SICK) {People per Week}
BETA = 0.000002 { + SINWAVE(0.0000005,52) }
POPULATION = NONIMMUNE + CONTAGIOUS + IMMUNE + SICK {People}
UNAWARE_CONTAGIOUS = CONTAGIOUS/3 {A third of the contagious persons are unaware of being contagious; People}
BIRTH_RATE = GRAPH(POPULATION)
(0.00, 0.00), (166667, 0.0115), (333333, 0.023), (500000, 0.0325), (666667, 0.041), (833333, 0.0525), (1000000, 0.061), (1.2e + 06, 0.0705), (1.3e + 06, 0.076), (1.5e + 06, 0.076), (1.7e + 06, 0.0735), (1.8e + 06, 0.066), (2e + 06, 0.045)
NAT_DEATH_RATE = GRAPH(POPULATION)
(0.00, 0.0005), (208333, 0.005), (416667, 0.0102), (625000, 0.0153), (833333, 0.0204), (1e + 06, 0.0249), (1.2e + 06, 0.0294), (1.5e + 06, 0.033), (1.7e + 06, 0.0363), (1.9e + 06, 0.0387), (2.1e + 06, 0.0411), (2.3e + 06, 0.0426), (2.5e + 06, 0.0432)

2.6 Questions and Tasks

1. Try running the models of Sections 2.1 and 2.2 with different time steps and explain why the results differ.

2. Similarly to the chemistry models using the law of mass action, you may want to change the "reaction rate" by, e.g., introducing an exponent α such that SICK RATE = CONTACT RATE * (CONTAGIOUS+SICK) *NON IM-MUNE. Vary α in consecutive runs, for example, set $\alpha = 0.9$, $\alpha = 1.1$, $\alpha = 1.3$. How do the model results differ? (Presumably, the basic form $\alpha = 1.0$ has shown some historical veracity; i.e., the form has been sufficiently fit with historical data.)

3. For the model of Section 2.1, connect the birth rate with the immune population and try to reach a steady-state immune population.

4. Does the disease die out of the population?

5. Is it possible to wipe out the population with a variation in the parameters in this form of the model?

6. Does your answer change if you allow for randomness of the contact rate?

7. Suppose that it takes anywhere between 1 and 5 days for someone to get so sick that they choose to stay in bed. Should that variation be reflected in the model as a random variable or be changed from model run to run?

8. Can you introduce an optimum (minimum number of sick) vaccination program to stabilize the disease in this latter form of the model?

9. Can you model how the disease can frustrate the vaccination program through mutation?

10. a) Can you break the population into age groups with different contact rates, death rates, birth rates, initial populations, and disease-induced death rates for each?
    b) Show how some of these folks seem to be more resistant to the disease and skip from contagious to immune directly.
    c) Show how the result changes when immunity is slowly lost. (Note that in reality, the immunes mingle with the nonimmunes and therefore dilute the original effect of the contact rate coefficient. Can you fix this problem?)

11. Develop an epidemics model that captures the same population but distinguishes two regions. Immune and contagious people can travel but sick individuals cannot. People from the two regions have different contact rates and are affected by the disease differently, i.e. the recovery rate differs between the two subgroups of the population. What are the implications for an optimal vaccination program that does not restrict travel between the regions?

12. Let the virus in the model of Section 2.4 evolve to a less deadly form in humans and show the impacts of this evolution.

13. Introduce a spatial component into the model of Section 2.5 by considering two regions with different contact rates and different vaccination programs. Investigate the implications of travel restrictions imposed by one of the regions on individuals originating in the other region.

14. Introduce longer delays in the CONTAGIOUS and SICK populations and elaborate on the effects of each on the trajectory of the IMMUNE population.

15. Show how the disease could be eliminated by reducing the CONTACT RATE 1_2 in the model of Section 2.5.

Chapter 3
Insect Dynamics

In this chapter we develop three different models of insect dynamics. From the perspective of "dynamic modeling of pestilence," insects surely need special attention—they are the carriers of many of the viruses that affect the health of humans and other animals, and they cause billions of dollars of damage to food supplies around the world every year. We try to control their population levels, having long ago realized they multiply and evolve too fast for elimination.

Insects are among the major vectors of disease. They provide an excellent way to understand the dynamic modeling of a complicated disease vector composed of different age classes.

From a modeling perspective, insect dynamics can be used to effectively illustrate a set of techniques and skills that will prove important for gaining a deeper understanding of

(a) the challenges of matching laboratory and field experiments to models,
(b) the underlying mechanisms by which living systems evolved and adjust to their external environment,
(c) the distinct roles that different life stages or age cohorts may play in the transmission of, or infection with a disease.

The following three sections of this chapter respectively address these issues. Later chapters either explicitly or implicitly make use of the insights gained here.

3.1 Matching Experiments and Models of Insect Life Cycles

To better understand the dynamics of insect populations, we model the life cycle of an insect, simplified into two stages, egg and adult. Typically, the data used in understanding insect population dynamics come from laboratory experiments in which one watches each egg and notes when it dies or hatches. Data from such an experiment (at constant temperature) might look like that shown in Table 3.1 of the life history of 100 new insect eggs. Note how the final number of survivors must equal the total number hatched.

B. Hannon and M. Ruth, *Dynamic Modeling of Diseases and Pests*,
Modeling Dynamic Systems,
© Springer Science + Business Media LLC 2009

Table 3.1

Time	Died	Survived	Hatched	Time*Survived	Time*Hatched
0	1	99	0	0	0
1	3	96	0	96	0
2	3	93	2	186	4
3	4	89	3	267	9
4	5	84	21	336	84
5	5	79	9	395	45
6	6	73	7	438	42
7	6	67	5	469	35
8	7	60	3	480	24
9	8	52	2	468	18
Col. Sum = 45			52	3135	261
				3135/45/100 =	261/52 =
				0.697 = ESF	5.02 = T

Time is measured in days in this case, with data displayed for the beginning of the next day (the result of the previous day). This table yields two important averaged numbers, the experimental survival fraction, ESF (0.699, say 0.7), and the experimental maturation time, T, (5.019, say 5 days). How can we use such data to parameterize a model, when the model time step is dramatically different from this experimentally found maturation time?

We must develop a new concept: the model survival fraction, MSF. In ecological experiments, the instantaneous survival rate cannot be measured. But the survival rate can be measured over some real time period, the maturation time, by counting the number of eggs surviving to maturation. A problem arises when we wish to model the system at a shorter time step than the real one. We need to model at these shorter times because the characteristic time of the system may be shorter than the shortest feasible measurement time of some particular part of the system. So we have the experimental time step (the maturation time) and the model time step (DT) and we must devise a conversion from experiment to model.

That conversion is based on the assumption that the survival fraction is a declining exponential, with ESF and T as its mean point.[1]

$$\text{ESF(t)} = N(t+DT)/N(t) = \text{EXP}(-m*t) \tag{3.1}$$

$$= \text{the dimensionless experimental survival fraction}$$

$$\text{as a function of time.}$$

N is the population size. Later, when we attempt to confirm the experimental data with our model, we will shut off the birth and hatch rate and observe the (necessarily exponential) decline in egg population due to death. The resulting instantaneous

---

[1] We are constrained here to the assumption of exponential decline since both the DEATH and HATCH flows in the model are arranged as exponential decays.

survival fraction is determined with the constant m. Using the experimental data, we solve this equation for −m:

$$-m = LOGN(ESF)/T. \tag{3.2}$$

The *model* survival fraction (based on our choice of the time step DT) is derived as follows:

$$MSF = ESF(t + DT)/ESF(t) = EXP(-m*DT). \tag{3.3}$$

When the expression for −m is substituted into (3.3), we get

$$MSF = EXP(LOGN(ESF)/T*DT), \tag{3.4}$$

which is the basic equation for the model survival fraction. We now have the instantaneous survival fraction, and those surviving will mature or hatch at the maturation or HATCH rate

$$HATCHING = EGGS/T*MSF, \tag{3.5}$$

that is, the survivors hatch at the rate, EGGS/T. Remember, eggs do not have to hatch or die. They may simply wait. When they do die, they are claimed at the model death rate

$$DYING = EGGS*(1 - MSF)/DT, \tag{3.6}$$

the multiplier fraction being the model death rate per egg.

Such life table data can be used to determine the death rate of the adults, a normal demographic application. If we were to watch 100 new adults, we could calculate the experimental adult survival fraction, EASF, and adult survival time (mean length of adult life), TA. Let us say that we found these numbers to be 0.8 and 1.0, respectively. These numbers are used in a parallel way to obtain the equivalent of equation 3.6 for ADULTS.

The layout for the egg–adult model is shown in Figure 3.1, with an EGG LAY RATE of 0.5 (eggs per adult per day) and the model results are shown in the graph of Figure 3.2. We realize that insect egg laying rate is not constant but either declining with time or pulsed. These concepts could be incorporated in more sophisticated versions.

Now turn off the BIRTHING and HATCHING flows and put 100 eggs in EGGS. Run the model to verify that it reproduces the experimental mean maturation rate, T, and the experimental survival fraction at T. This must be so, since our modeling process is one of exponential decay for both the DYING and HATCHING flows.

TWO STAGE INSECT POPULATION MODEL

ADULTS(t) = ADULTS(t − dt) + (HATCHING − ADULTS_DYING) * dt
INIT ADULTS = 0
INFLOWS:
HATCHING = EGGS/T*MSF
OUTFLOWS:
ADULTS_DYING = ADULTS*(1 − MASF)/DT

EGGS(t) = EGGS(t − dt) + (BIRTHING − EGGS_DYING − HATCHING) * dt
INIT EGGS = 50
INFLOWS:
BIRTHING = EGG_LAY_RATE*ADULTS
OUTFLOWS:
EGGS_DYING = EGGS * (1 − MSF)/DT
HATCHING = EGGS/T * MSF

EASF = .8* EGG_LAY_RATE = .5
ESF = .7
MASF = EXP(LOGN(EASF) * DT/TA)
MSF = EXP(LOGN(ESF)/T * DT)
T = 5
TA = 1

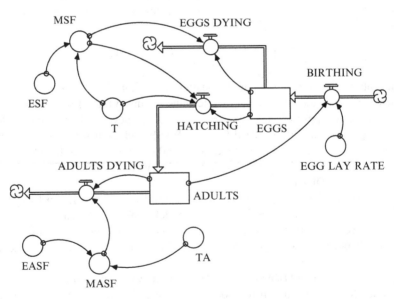

Fig. 3.1

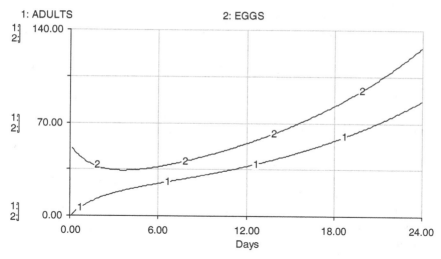

Fig. 3.2

3.2 Optimal Insect Switching

Few attempts are made to consider anything in nature other than humans as be-
having in some optimal way. We are loathe to consider any kind of optimal
behavior in plants or animals since we think that such behavior requires intelligence... intelligence of the kind we believed to be the sole domain of humans.

We need not think of intelligence as a prerequisite of optimal behavior at all.
It is well known in engineering that the configuration of a structural member can
be accurately calculated from a theory stating that the member deforms in such a
way as to minimize its internal strain energy. A steel beam can do this, without
knowledge of differential equations or modeling!

Yet in some real sense, we accept that evolution produces certain forms of efficient behavior. The idea that energy is most efficiently used in nature is an example.
We can find such efficiencies in the field. Following millennia of evolution, a plant
possesses an optimal switching time between time spent growing roots, stems, and
leaves, and time spent growing seed. We can imagine that the plant that behaves this
way would have the most offspring in the next generation and so be the most likely
plant present in future times. But it is surely not only the amount of offspring that a
species produces that determines their prevalence in the future. The care with which
each offspring is produced and engendered must have some effect as well. Think of
the oak tree compared to the maple. The oak produces relatively few, well-packaged
seeds while the maple sends thousands of relatively unclad seeds into the wind. Here
we examine one species in our search for the first indications of an optimal strategy.

Precisely what are we suggesting when we say that a plant or an animal switches
optimally? We are really admitting that such behavior does not require intelligence
of any kind. We are saying that evolution favors efficient strategies. Our problem

is to figure out what that strategy is and how it operates. We are able to show that some plants act as though they know when to optimally switch from vegetation to seed production[2]. Here we assume that insects might employ a similar kind of optimal strategy in switching egg maturation times to maximize the number of eggs produced.

Consider two forms of a particular insect whose life cycle was described in Chapter 3.1: EGGS and ADULTS. As before, let them each have an experimental survival fraction and an average maturation time and let the ADULTS have a particular egg-laying rate. Start the model with no adults and 50 eggs. Now suppose that the life span of these adults was only 24 days and suppose that they could somehow adjust the egg average maturation time. Think of the insect species as originally having a variety of maturation rate responses to a given temperature. Assume that the most successful insect subspecies is the one whose average egg maturation time maximizes the number of eggs at the end of the 24-day adult life period. What is this optimum average egg maturation time?

We set up a sensitivity analysis and vary T to find the largest egg clutch in 24 days. The optimal average egg maturation time for our imagined insect species is 1.54 days (unrealistically short, except perhaps for certain mosquito species), as one can find in the table with the actual model. That such an optimal T exists for this model when the remaining parameters are fixed leads us to ask if that species has evolved to an optimal T. We would be assuming that the model structure was accurate and that the parameters were accurately measured. If not, then our model could be wrong or the insect might have another goal. In any case, it should be clear that such a model allows us to address very interesting questions about the insect, questions that could not be framed without such modeling.

Still another value of the average maturation time may someday be found in a model. Suppose that an insect was a link in a disease transmission. We could study the effect of introducing genetically modified insects such that the disease link is broken by extended or shortened average maturation time. In this way, the modified insect would occupy the critical ecological niche but produce an under- or over-supply of eggs, restricting the spread of the disease.

3.3 Two-Age Class Parasite Model

Now that we have modeled in more detail the dynamics of an insect population, let us turn to the spread of disease within different cohorts of an insect population, such as asexually reproducing aphids, consisting of two life stages—nymphs

---

[2] Cohen, D. 1971. Maximizing final yield when growth is limited by time or by limiting resources, *J. Theo. Biol.* 33, 299–307.

Chiariello, N. and J. Roughgarden. 1984. Storage allocation in seasonal races of an annual plant: optimal vs. actual allocation, *Ecology* 65-4, 1290–1301.

Kozlowski, J. and R. Wiegert. 1986. Optimal allocation of energy to growth and reproduction, Theo. Pop. Biol. 29, 16–37.

Hannon, B. 1993. The optimal growth of Helianthus Annus, J. Theo. Biol. 165, 523–531.

and adults. The model has two parts, one for the healthy population and one for the diseased population. Diseased nymphs can infect healthy nymphs and become diseased adults. Diseased adults cannot infect healthy adults or nymphs but can produce infected nymphs. Note how these features are expressed in the model by the appropriate flows and links.

Assume that the infection coefficient is based on an exponential model:

$$\text{INFECTION COEF} = 1 - \text{EXP}(-.3 * \text{NYMPHS} * \text{NYMPHS D}) \qquad (3.7)$$

NYMPHS and NYMPHS D refer to the population sizes of healthy and infected nymph populations, respectively. The INFECTION RATE is calculated as the product of the INFECTION COEF, the number of healthy nymphs, divided by the model maturation rate for survivors, MNSF/TN

$$\text{INFECTION RATE} = \text{MNSF/TN} * \text{INFECTION COEF} * \text{NYMPHS} \qquad (3.8)$$

with

$$\begin{aligned} \text{MNSF} &= \text{Model Nymph Survival Fraction} \\ &= \text{EXP(LOGN(ENSF)/TN} * \text{DT)} \end{aligned} \qquad (3.9)$$

where TN is the experimental nymph maturation rate and ENSF is the dimensionless experimental nymph survival fraction. When there are no sick nymphs or healthy nymphs, the probability of becoming infected equals zero. The specification of the INFECTION COEF translation variable is a purely empirical formulation but it gives the correct value at the extremes: 0 when the number of diseased nymphs is zero or when the number of healthy nymphs is zero, and near 1 when either at least one of the stocks NYMPHS or NYMPHS D is very large.

Note well the specification of the MATURING function in the model, which ensures that the total rate of change from nymphs to adults here is still U1*NYMPHS:

$$\text{MATURE} = \text{MNSF/TN} * \text{NYMPHS} * (1 - \text{INFECTION COEF}) \qquad (3.10)$$

Figure 3.4 shows the proportions of healthy nymphs and adults, and the number of diseased nymphs and adults. Similar to the previous section of this chapter, we find distinct phases for the outbreak of a disease.

These newly diseased nymphs are converted to diseased adults rather than directly into diseased nymphs in an effort to reflect the fact that these nymphs, who contract rather than acquire the disease, have the normal nymph survival rate. In addition, they are unable to convey the disease to other healthy nymphs.

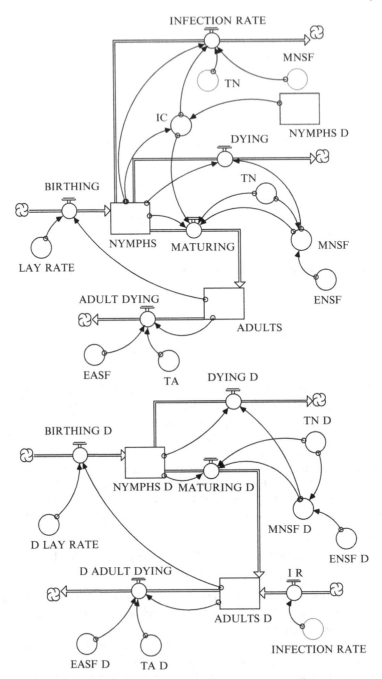

Fig. 3.3

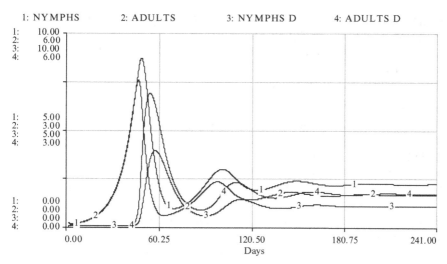

1: NYMPHS   2: ADULTS   3: NYMPHS D   4: ADULTS D

1:	10.00
2:	6.00
3:	10.00
4:	6.00

1:	5.00
2:	3.00
3:	5.00
4:	3.00

1:	0.00
2:	0.00
3:	0.00
4:	0.00

0.00   60.25   120.50   180.75   241.00
Days

Fig. 3.4

TWO-AGE CLASS PARASITE MODEL

ADULTS(t) = ADULTS(t − dt) + (MATURING − ADULT_DYING) * dt
INIT ADULTS = .10 {Individuals}

INFLOWS:
MATURING = MNSF * NYMPHS*(1 − IC)/TN {Individuals per Day}
OUTFLOWS:
ADULT_DYING = ADULTS * (1 − EXP(LOGN(EASF)/TA * DT))/DT
{Individuals per Day}
ADULTS_D(t) = ADULTS_D(t − dt) + (MATURING_D + I_R −
D_ADULT_DYING) * dt
INIT ADULTS_D = .10 {Individuals}

INFLOWS:
MATURING_D = MNSF_D*NYMPHS_D/TN_D
I_R = INFECTION_RATE {Individuals per Day}
OUTFLOWS:
D_ADULT_DYING = ADULTS_D * (1 − EXP(LOGN(EASF_D)/TA_D *
DT))/DT {Individuals per Day}
NYMPHS(t) = NYMPHS(t − dt) + (BIRTHING − DYING − MATURING −
INFECTION_RATE) * dt
INIT NYMPHS = 0 {Individuals}

INFLOWS:
BIRTHING = LAY_RATE*ADULTS {Individuals per Day}
OUTFLOWS:
DYING = (1 − MNSF)*NYMPHS/DT

MATURING = MNSF * NYMPHS*(1 − IC)/TN {Individuals per Day}
INFECTION_RATE = MNSF*IC*NYMPHS/TN
NYMPHS_D(t) = NYMPHS_D(t − dt) + (BIRTHING_D − DYING_D −
MATURING_D) * dt
INIT NYMPHS_D = 0 {Individuals}

INFLOWS:
BIRTHING_D = D_LAY_RATE * ADULTS_D {Individuals per Day}
OUTFLOWS:
DYING_D = (1 − MNSF_D)*NYMPHS_D/DT
MATURING_D = MNSF_D * NYMPHS_D/TN_D
D_LAY_RATE = .35 {Experimental laying fraction. Eggs per Adult per Day}
EASF = .8
EASF_D = .65 {Experimental daily adult survival fraction per stage,
dimensionless.}
ENSF = .7 {Experimental egg survival fraction, dimensionless, per stage. Stage
= 1/F1, i.e., 70 eggs per 100 eggs survive each 1/F1 days, as noted in the
experiment.}
ENSF_D = .5 {Experimental egg survival fraction, dimensionless, per stage.
Stage = 1/F1, i.e., 70 eggs per 100 eggs survive each 1/F1 days, as noted in the
experiment.}
IC = 1 − EXP(−.3*NYMPHS*NYMPHS_D)
DOCUMENT: INFECTION COEFFICIENT

LAY_RATE = .6 {Experimental laying fraction. Eggs per Adult per Day}
MNSF = EXP(LOGN(ENSF)/TN*DT)
MNSF_D = EXP(LOGN(ENSF_D)/TN_D * DT)
TA = 1
TA_D = 1
TN = 5
TN_D = 5

3.4 Questions and Tasks

1. Suppose we are uncertain about the exact egg experimental fraction in the model
   of Section 3.2. We may suspect that using literature data is not good enough, and
   think that this number is within ±10%. We can conduct an experiment to find
   this number if the total number of adults in 24 days is within ±10%.
2. Insert a larval stage into the model of Section 3.2 with a larval survival fraction
   of 0.8 in 3 days maturation time. Why does the stock of adults in this model fail
   to grow as is did in the previous version?
3. (a) For the model in Section 3.2, change the egg-laying rate from 0.5 to 0.45 and
   find the new optimal egg-laying rate.

(b) Are any of the other parameters in this model susceptible to such an optimal-
ity analysis, that is, do any of the other parameters show any particular value
other than an extreme that leads to a maximum number of eggs at the end of
the season?

4. For the model in Section 3.3, vary the separate birth and death rates and the
infection coefficient and note the effect on the relative size of the healthy and
diseased portions of the populations. Can you identify some relationship between
these parameters that enable you to judge the effects on the relative size of the
healthy and diseased portions of the populations?

Part II
Applications

Chapter 4
Malaria and Sickle Cell Anemia

4.1 Malaria

4.1.1 Basic Malaria Model

Malaria is one of the most severe human diseases, causing more than 300–500 million cases today[1], leading to an estimated 2.7 million deaths worldwide with 80–90% of those occurring in the section of Africa below the Sahara Desert. Children ages one to four are most vulnerable to malaria due to their immature immune systems. Malaria is caused by a parasite transmitted to humans or animals by the *Anopheles* mosquito. The human parasite, *Plasmodium falciparum*, digests the hemoglobin found in red blood cells (RBCs) and breaks down the adhesive properties of the cells. Therefore the RBCs may become stuck to the walls of capillaries. When this occurs in the cerebral section of the brain, cerebral malaria develops, and blood clots in the brain occlude the vessels. Symptoms of malaria increase in severity each time an RBC bursts and releases more parasites. These symptoms can include high fever, chills, and uncontrollable shaking/shivering, also called rigors. Severe headache, vomiting, muscle pain, and extreme tiredness may accompany these symptoms as well. To combat the disease, many countries use an insecticide called DDT to control mosquito populations. Different medications have also been developed to treat and prevent malaria. These include chloroquine, doxycycline, and mefloquine.

Several variables are implicated in the epidemiology of this disease. In this chapter we develop a simplified model to investigate the dynamics of the spread of malaria in a closed ideal region without human immigration. We further assume that humans are the most important hosts and disregard other possible hosts aside from the mosquitoes, which pick up the malaria parasites when they feed on the blood of an infected human. Upon receipt of malaria parasites (*Plasmodium sp.*),

---

[1] Martens, P., R.S. Kovats, S. Nijhof, P. de Vries, M.T.J. Livermore, D.L. Bradley, J. Cox, and A.J. Mitchel. 1999. Climate change and future populations at risk of malaria. Global Environmental Change. 9:S89–S107.

B. Hannon and M. Ruth, *Dynamic Modeling of Diseases and Pests*,
Modeling Dynamic Systems,
© Springer Science + Business Media LLC 2009

Fig. 4.1

the parasites reproduce in the gut of mosquitoes. Subsequently, malaria parasites are returned through the mosquitoes' salivary glands to humans as the mosquitoes feed on red blood cells.

The mosquito hosts cannot be just "any" mosquito but must be of the genus *Anopheles*. Many species of mosquito in the genus *Anopheles* are vectors, and we will use information on the one that is the most widespread in South and Central America, *A. messae*[2]. Among the four species of parasites, we will concentrate on *Plasmodium vivax*; this species presents the most extensive geographical spread and is present both in tropical and temperate zones.

The STELLA model of the malaria life cycle is divided into several interrelated modules, which we visually separated into "sectors." Sectors enable us to organize parts of a model by grouping model components. They also add functionality, as they allow entities in a sector to be moved as a whole, or run separately from the rest of a model. To create a sector, choose the Sector Frame among STELLA's tools (Figure 4.1), place it in the model diagram, and drag it over those components of a model that you wish to group together. Clicking on the "lock" in the upper right hand side of a sector fixes all elements in that sector so you can move them as a whole. Once you have sectors specified, the Run pull-down menu gives you the option of running an entire model or specific sectors. If you run sectors individually, then values of parameters from other sectors, which do not run, will be assumed as given and fixed.

All of the modules of this chapter's model are grouped into individual sectors, mainly to keep the model organized and its structure transparent. One of the modules is used to calculate reproduction of parasites (Figure 4.2). This module contains the various parameters affecting the basic reproductive rate, Ro, which is modeled here as the average number of new cases of the disease that would arise from a single infectious host introduced into a population of susceptible hosts. It is a function of the probability of transmission of parasites from infected vertebrates to uninfected vertebrates (TR), the proportion of vectors to human hosts (M) and the vector daily survival rate[3,4]:

$$\text{Ro} = (\text{TR} * \text{M})/(-\text{LOGN}(\text{DAILY\_SURVIVAL\_VECTOR}))\qquad(4.1)$$

[2] Jetten, T.H. and W. Takken. 1994. Anophelism with malaria in Europe: a review of the ecology and distribution of the genus. *Anopheles* in Europe. Wageningen Agricultural University Press, Wageningen.

[3] Anderson, R.M. and R.M. May. 1991. Infectious diseases of humans: dynamics and control. Oxford University Press, Oxford.

[4] Begon, M., J. Harper, and C.R. Townsend. 1996. ECOLOGY. 3rd edition.

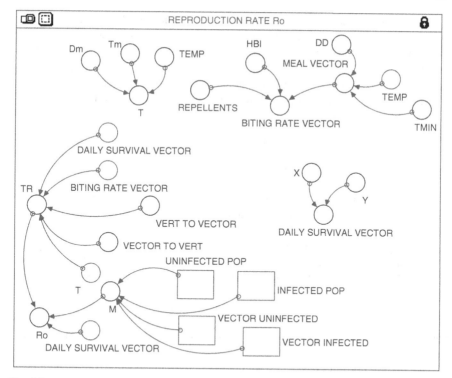

Fig. 4.2

where the transmission rate TR is defined as

$$\text{TR} = (\text{BITING\_RATE\_VECTOR}^{\wedge}2) * \text{VERT\_TO\_VECTOR} \\ * \text{VECTOR\_TO\_VERT} * (\text{DAILY\_SURVIVAL\_VECTOR}^{\wedge}\text{T}) \tag{4.2}$$

where VERT TO VECTOR and VECTOR TO VERT are, respectively, the probability of transmission of parasites from infected vertebrates to uninfected vectors, and from vectors to vertebrates[5]. For the model, these probabilities are assumed to be fixed. In reality, though, these probabilities depend on absolute and relative vector and vertebrate population densities.

M is the proportion of vectors to human hosts:

$$\text{M} = (\text{VECTOR\_INFECTED} + \text{VECTOR\_UNINFECTED})/ \\ (\text{INFECTED\_POP} + \text{UNINFECTED\_POP}) \tag{4.3}$$

The mosquito daily biting rate, in turn, is

$$\text{BITINGRATEVECTOR} = \text{HBI} * \text{MEAL\_VECTOR}/\text{REPELLENTS} \tag{4.4}$$

---

[5] Rogers, D.J. and S.E. Randolph. 2000. The global spread of malaria in a future, warmer world. *Science* 289:1763–1766.

We assume that the rate at which mosquitoes bite humans (BITING RATE VEC-TOR) is the product of the frequency with which a vector takes a blood meal (MEAL VECTOR) and the proportion of these meals that are taken from humans—the "human blood index" (HBI in the model). Following,[1] values for HBI may lie in the range from 0.005 to 0.63. By choosing alternative values for the converter REPEL-LENTS, contact among vectors with humans, and thus the rate of transmission of the disease, can be changed.

The MEAL VECTOR depends on temperature,

$$\text{MEALVECTOR} = D/(\text{TEMP} - \text{Tmin}), \tag{4.5}$$

where D is the number of the degree days required for the completion of development (36.5 days), Tmin is a minimum temperature requirement for parasite development (9.9 degrees C), and TEMP is average assumed temperature for the tropics.[1]

Similarly, the length of incubation T of parasites in vectors, measured in days, is a function of the temperature threshold Tmin, a minimum number of degree days required for development, Dm, and average temperatures.

$$T = Dm/(\text{TEMP} - \text{Tm}). \tag{4.6}$$

Following[6] we assume that 105 degree days are needed for the development of the parasite.

The female mosquito has to live long enough for the parasite to complete its development if transmission is to occur. Longevity of the mosquito vector depends on the species, humidity, and availability of hosts and temperature. Actual transmission intensity also depends on vector abundance, which is not modeled here. The daily survival of vectors is calculated from the survival probability, Y, during one gonothrophic cycle, and length of the cycle, $X^{(1,3)}$ as

$$\text{DAILYSURVIVALVEC} = X^{\wedge}(1/Y) \tag{4.7}$$

The second module (Figure 4.3) focuses on the spread of the infection in the vectors. To be infectious, an uninfected vector must get the parasite from a human host. The development of the parasite in the vector requires certain temperatures included in the model. Most of the steps followed in this module of vector birthing, hatching, and dying are based on the insect life stages model developed in[7].

The third module (Figure 4.4) uses, for illustrative purposes, population numbers from Venezuela (U.S. Census Bureau) as well as data from the literature to simulate the recovery rate of infected people and their loss of immunity. This module follows the structure of the basic epidemic models developed in Chapter 2. Loss of immunity is related to the transmission rate. If the transmission rate is high and humans are still

---

[6] MacDonald, G. 1957. The epidemiology and control of malaria. Oxford University Press, London, UK.

[7] Hannon, B. and M. Ruth. 1997. Modeling dynamic biological systems. Springer-Verlag, New York.

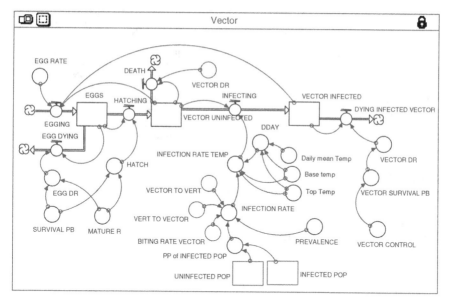

Fig. 4.3

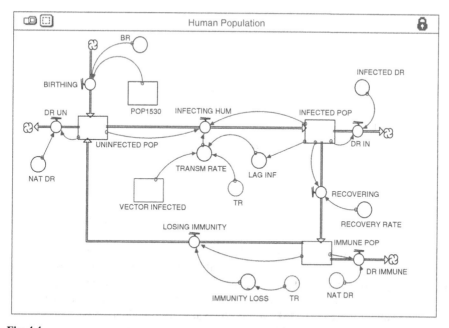

Fig. 4.4

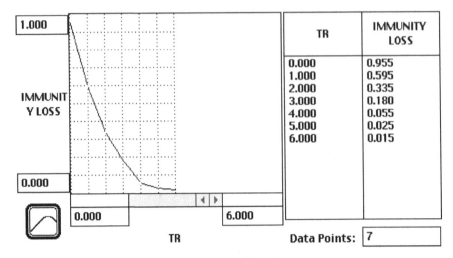

TR	IMMUNITY LOSS
0.000	0.955
1.000	0.595
2.000	0.335
3.000	0.180
4.000	0.055
5.000	0.025
6.000	0.015

Data Points: 7

Fig. 4.5

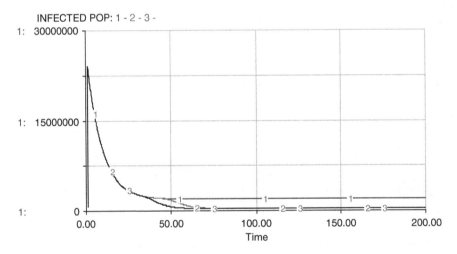

Fig. 4.6

being bitten, their loss of immunity is low (Figure 4.5). The death rate of infected people is higher than the natural rate, especially among children. We use an average death rate for different cohorts to calculate the death rate of infected people.

Let us now explore a set of scenarios for this model. In the first scenario, we assume a value for HBI = 0.0005 and set the value for repellents = 10. This means a small biting rate vector. For this, we change the vector control between 0.1 (very low control of vectors, and thus high survival rate of vectors) and 1.2 (high control of vectors, and thus low survival rate of vectors). The results are shown in Figure 4.6

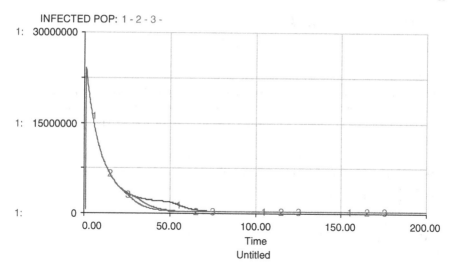

Fig. 4.7

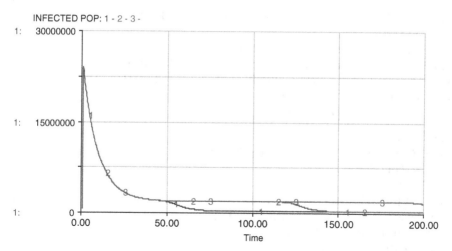

Fig. 4.8

and indicate, as expected, a more prolonged persistence of malaria (a larger infected population) in the long run, when controls are low. Similarly, more extensive use of repellents reduces human exposure and thus the size of the infected population. This is shown in Figure 4.7, where HBI = 0.0005, Vector Control = 1.0, and the variable Repellents increases from 10 to 30, and finally to 50 for the last run.

The last set of model runs assume Repellents = 10, Vector Control = 1.0 and HBI increases from run to run by an order of magnitude, starting with 0.0005 and ending with 0.05. The results are shown in Figure 4.8. Higher HBI yield, in the long run, larger infected populations.

4.1.2 Questions and Tasks

1. How would the behavior of the model change if we include resistance of the vectors against repellents and insecticides?
2. How can we include the influence of socioeconomic factors into the formula, such as increased education about causes and transmission of malaria, and induced behavioral changes, or improved health care (such as expanded availability of drugs)?
3. How would the spread of the disease be affected by a global climatic change?
4. How may rains affect the spread of the disease?

BASIC MALARIA MODEL

Human Population
IMMUNE_POP(t) = IMMUNE_POP(t − dt) + (RECOVERING −
DR_IMMUNE − LOSING_IMMUNITY) * dt
INIT IMMUNE_POP = 0

INFLOWS:
RECOVERING = RECOVERY_RATE*INFECTED_POP
OUTFLOWS:
DR_IMMUNE = IMMUNE_POP*NAT_DR
LOSING_IMMUNITY = IMMUNITY_LOSS * IMMUNE_POP
INFECTED_POP(t) = INFECTED_POP(t − dt) + (INFECTING_HUM −
RECOVERING − DR_IN) * dt
INIT INFECTED_POP = 377000

INFLOWS:
INFECTING_HUM = TRANSM_RATE*UNINFECTED_POP *
INFECTED_POP
OUTFLOWS:
RECOVERING = RECOVERY_RATE * INFECTED_POP
DR_IN = INFECTED_POP * INFECTED_DR
POP1530(t) = POP1530(t − dt)
INIT POP1530 = 8676000

UNINFECTED_POP(t) = UNINFECTED_POP(t − dt) +
(LOSING_IMMUNITY + BIRTHING − DR_UN − INFECTING_HUM) * dt
INIT UNINFECTED_POP = 23706000

INFLOWS:
LOSING_IMMUNITY = IMMUNITY_LOSS * IMMUNE_POP
BIRTHING = POP1530 * BR
OUTFLOWS:
DR_UN = UNINFECTED_POP * NAT_DR
INFECTING_HUM = TRANSM_RATE*UNINFECTED_POP *
INFECTED_POP
BR = 0.021

INFECTED_DR = 0.1127
LAG_INF = DELAY(INFECTED_POP, 0.03)
NAT_DR = 0.005
RECOVERY_RATE = 0.0125
TRANSM_RATE = VECTOR_INFECTED*LAG_INF * TR
IMMUNITY_LOSS = GRAPH(TR)
(0.00, 0.955), (1.00, 0.595), (2.00, 0.335), (3.00, 0.18), (4.00, 0.055), (5.00, 0.025), (6.00, 0.015)

REPRODUCTION RATE Ro
BITING_RATE_VECTOR = HBI * MEAL_VECTOR/REPELLENTS
DAILY_SURVIVAL_VECTOR = $X^{(1/Y)}$
DD = 36.5
Dm = 105
HBI = .05
M = (VECTOR_INFECTED+VECTOR_UNINFECTED)/(INFECTED_POP + UNINFECTED_POP)
MEAL_VECTOR = DD/(TEMP-TMIN)
REPELLENTS = 10
Ro = (TR * M)/(−LOGN(DAILY_SURVIVAL_VECTOR))
T = Dm/(TEMP − Tm)
TEMP = 25
Tm = 15
TMIN = 9.9
TR = $(BITING\_RATE\_VECTOR^2)$ * VERT_TO_VECTOR * VECTOR_TO_VERT * $(DAILY\_SURVIVAL\_VECTOR^T)$
VECTOR_TO_VERT = 1
VERT_TO_VECTOR = 1
X = 0.61
Y = 3

Vector
EGGS(t) = EGGS(t − dt) + (EGGING − HATCHING − EGG_DYING) * dt
INIT EGGS = 1000

INFLOWS:
EGGING = EGG_RATE * (VECTOR_UNINFECTED + VECTOR_INFECTED)
OUTFLOWS:
HATCHING = EGGS * HATCH
EGG_DYING = EGGS * EGG_DR
VECTOR_INFECTED(t) = VECTOR_INFECTED(t − dt) + (INFECTING − DYING_INFECTED_VECTOR) * dt
INIT VECTOR_INFECTED = 1000

INFLOWS:
INFECTING = INFECTION_RATE_TEMP * VECTOR_UNINFECTED

OUTFLOWS:
DYING_INFECTED_VECTOR = VECTOR_DR * VECTOR_INFECTED
VECTOR_UNINFECTED(t) = VECTOR_UNINFECTED(t − dt) +
(HATCHING − DEATH − INFECTING) * dt
INIT VECTOR_UNINFECTED = 1500

INFLOWS:
HATCHING = EGGS * HATCH
OUTFLOWS:
DEATH = VECTOR_DR * VECTOR_UNINFECTED
INFECTING = INFECTION_RATE_TEMP * VECTOR_UNINFECTED
Base_temp = 15
Daily_mean_Temp = 25 + 15 * SIN(2 * PI * TIME/12)
DDAY = if Daily_mean_Temp + 15 < Base_temp or Daily_mean_Temp + 15 >
Top_Temp then 0 else (Daily_mean_Temp + 15 − Base_temp)/2
EGG_DR = (1 − (EXP(LOGN(SURVIVAL_PB) * MATURE_R * DT)/DT))
EGG_RATE = 0.6
HATCH = MATURE_R * EXP(LOGN(SURVIVAL_PB) * DT * MATURE_R)
INFECTION_RATE = BITING_RATE_VECTOR * PP_of_INFECTED_POP *
VERT_TO_VECTOR * VECTOR_TO_VERT * PREVALENCE
INFECTION_RATE_TEMP = if DDAY > Base_temp or Base_temp < Top_Temp
then INFECTION_RATE else 0
MATURE_R = 0.2
PP_of_INFECTED_POP = INFECTED_POP/UNINFECTED_POP
PREVALENCE = 0.0186
SURVIVAL_PB = RANDOM(0.5, 0.7)
Top_Temp = 30
VECTOR_CONTROL = 1
VECTOR_DR = (1 − (EXP(LOGN(VECTOR_SURVIVAL_PB))))/DT
VECTOR_SURVIVAL_PB = Random(0.6,0.8)/VECTOR_CONTROL

4.2 Sickle Cell Anemia and Malaria in Balance

4.2.1 Sickle Cell Anemia

Sickle cell disease is an inherited disorder of the red blood cells. It is inherited as an autosomal recessive trait. Someone who inherits hemoglobin S from one parent and the normal hemoglobin A from the other parent will have the sickle cell trait. In an unaffected individual, red blood cells are doughnut-shaped and contain hemoglobin, the oxygen carrying protein. In a person who has sickle cell anemia, however, the hemoglobin is in the shape of a sickle. As a result, the cells function abnormally and cause small blood clots by sticking to each other on the walls of the bloodstream. These clots give rise to recurrent painful episodes called "sickle cell pain crises."

A few symptoms of sickle cell anemia are joint pain, fatigue, breathlessness, delayed growth, jaundice, abdominal pain, and susceptibility to infections.

Hemolytic crisis, sequestration crisis, and aplastic crisis are three different types of episodes. Hemolytic crisis is a result of damaged red blood cells that break down and is one of many life-threatening consequences of having sickle cell anemia. Sequestration crisis is due to the spleen enlarging and trapping the blood cells. Aplastic crisis is an infection that causes the bone marrow to stop producing red blood cells. Without proper medical attention, individuals with this disease often die at an early age.

Although no cure exists for this disease, a few steps can be taken to help prevent people from inheriting sickle cell anemia in the future. Prenatal diagnosis of sickle cell anemia and genetic counseling are available. As for the future fight against this disease, improvements continue to be made, and as the medical community continues to advance so will the health and lives of those with sickle cell anemia.

People with the sickle cell trait, a genetic condition, are not as easily infected with malaria. Heterozygous individuals having no or slight sickle cell anemia are more resistant to malaria because the parasite is unable to grow in red blood cells with sickle cell hemoglobin.

The following model mimics the balance between sickle cell anemia and malaria in developing African countries. We will use this model to explore the effects of antimalaria drugs and medical treatment on the frequency of sickle cell anemia. Specifically, the model focuses on the genetic balancing act between sickle cell anemia in a representative African population that has no immigration or emigration. Malaria infection is assumed to be largely endemic.

Infectious disease population model components are set up individually for the three possible genotypes (AA, AS, or SS), with susceptible (SUSC), infectious (INF), and immune stocks for each (Figures 4.9–4.11). The rate of infection is dependent on the number of susceptibles in the particular genotype, the total number of infectious individuals following a lag due to a latent period (A DELAY), and a transmission coefficient (Trans), which is a function of a transmission rate (TR) and the level of infected vectors.

The level of infected vectors is complex and could be modeled separately, but since we are modeling an endemically infected population, this parameter is simplified by providing only seasonal variability and a degree of randomness. The AS and SS genotypes have a lower transmission rate TR than the AA genotype since these individuals have partial resistance to malaria, unlike the AA individuals.

All of the stocks have a death rate associated with them. Most have a natural death rate, while the infectious stocks have a higher death rate due to malaria. Also, the SS genotype stocks have an additional death rate due to sickle cell anemia (Anemia DR). Individuals recover at a certain rate, and immunity is lost at a certain rate.

The total number of each genotype and allele, as well as the frequencies of each, are calculated in the genetic portion of the model (Figure 4.12). The allelic frequencies at each time step are used to determine the genotype of individuals born into the population and therefore which of the three diseased population models individuals will enter.

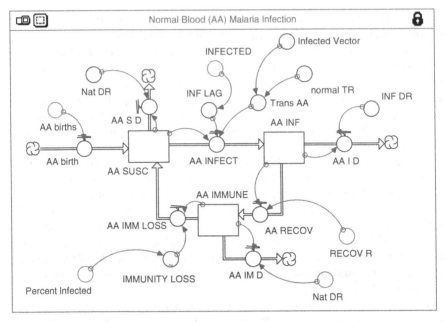

Fig. 4.9

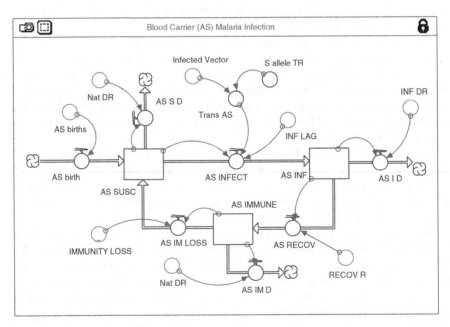

Fig. 4.10

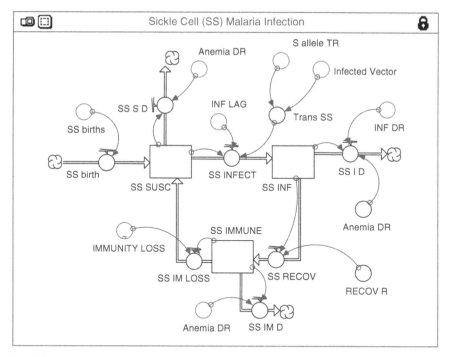

Fig. 4.11

The source for each parameter value is given in the document section of the parameter menus. After entering parameter values from the literature, the normal TR and S allele TR were determined empirically. First, we adjusted both equally to a value that caused epidemics of reasonable size before the disease becomes endemic as the model runs. When the two TRs were equal, the S allele frequency decreased over time and eventually approaches zero. Then we reduced the S allele TR to account for the fact that carriers of this allele are more resistant to malaria. This TR was adjusted to a value that resulted in 16% frequency of the S allele. With these empirical values, any initial stock values can be used, and the model will come to an equilibrium in the frequency of the S allele and the level of malaria infection.

Antimalaria drugs would decrease the death rate due to malaria and increase the recovery rate, so in the model, we changed the INF DR from 0.1127 to 0.05 and increased the RECOV R from 0.0125 to 0.1. This resulted in the S allelic frequency reducing over time rather than leveling off at 16%. After 1,000 cycles, the frequency is at 2.6%. This would be expected, since a reduced malaria infection level would no longer make the individuals with the AS genotype have an advantage over the individuals with the AA genotype (Figure 4.13). The corresponding numbers of susceptible, infected, and immune is shown in Figure 4.14.

If sickle cell anemia treatment becomes available, affected individuals will live longer and reproduce. If the Anemia DR is decreased, the equilibrium S allelic frequency increases. Figure 4.15 shows a sensitivity run with the Anemia DR

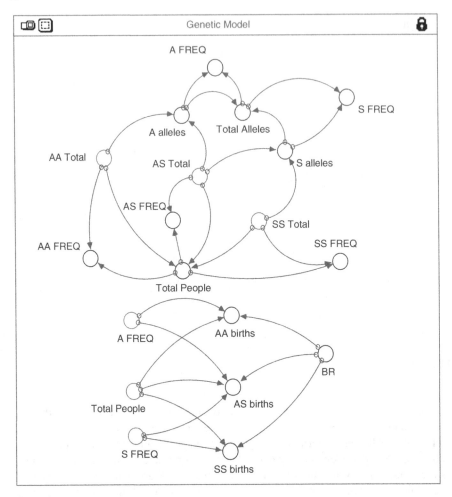

Fig. 4.12

decreasing from 0.5. Runs 1 through 5 have the following Anemia DRs: 0.5, 0.25, 0.1, 0.05, and 0.02. Since individuals with the S allele still have an advantage due to the malaria, as the Anemia DR approaches the natural DR, the S allelic frequency increases above 50%.

4.2.2 Questions and Tasks

1. How would you include the possibility for mutation in the model, and how would the prevalence of malaria and sickle cell anemia change in the light of mutation?

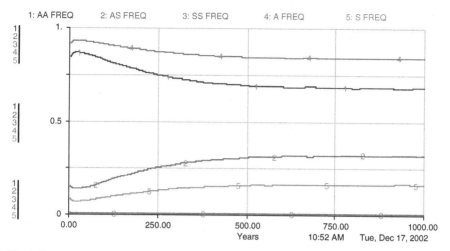

Fig. 4.13

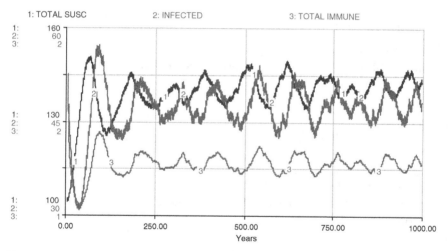

Fig. 4.14

2. In this model, how would you reflect the effects of mosquito repellents and malaria prophylaxis medication, which we investigated in Section 4.1?
3. Reformulate the model to reflect different age cohorts and introduce different age-specific recovery rates from malaria. What are the effects on the prevalence of malaria and sickle cell anemia?
4. If the vector and its infection level were included, would this affect the results? Would this allow the model to be used on other populations, such as those that do not have an endemic malaria problem?

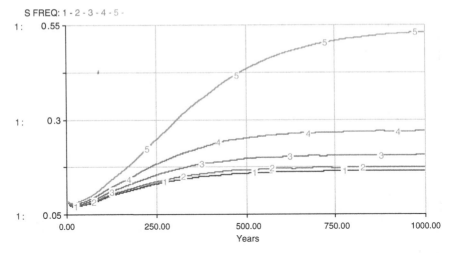

S FREQ: 1 - 2 - 3 - 4 - 5 -

Fig. 4.15

BALANCE OF MALARIA AND SICKLE CELL ANEMIA

Blood Carrier (AS) Malaria Infection
AS_IMMUNE(t) = AS_IMMUNE(t − dt) + (AS_RECOV − AS_IM_LOSS −
AS_IM_D) * dt
INIT AS_IMMUNE = 1

INFLOWS:
AS_RECOV = AS_INF*RECOV_R
OUTFLOWS:
AS_IM_LOSS = AS_IMMUNE*Immunity_Loss
AS_IM_D = AS_IMMUNE*Nat_DR
AS_INF(t) = AS_INF(t − dt) + (AS_INFECT − AS_RECOV − AS_I_D) * dt
INIT AS_INF = 12

INFLOWS:
AS_INFECT = AS_SUSC*Trans_AS * INF_LAG
OUTFLOWS:
AS_RECOV = AS_INF * RECOV_R
AS_I_D = AS_INF * INF_DR
AS_SUSC(t) = AS_SUSC(t − dt) + (AS_birth + AS_IM_LOSS − AS_INFECT
− AS_S_D) * dt
INIT AS_SUSC = 10

INFLOWS:
AS_birth = AS_births
AS_IM_LOSS = AS_IMMUNE * Immunity_Loss
OUTFLOWS:
AS_INFECT = AS_SUSC * Trans_AS * INF_LAG

AS_S_D = AS_SUSC * Nat_DR
S_allele_TR = 0.00066
Trans_AS = Infected_Vector * S_allele_TR

Conversions
AA_Total = AA_IMMUNE + AA_INF + AA_SUSC
AS_Total = AS_IMMUNE + AS_INF + AS_SUSC
INFECTED = AA_INF + AS_INF + SS_INF
Percent_Infected = INFECTED/Total_People
SS_Total = SS_IMMUNE + SS_INF + SS_SUSC
TOTAL_IMMUNE = AA_IMMUNE + AS_IMMUNE + SS_IMMUNE
TOTAL_SUSC = AA_SUSC + AS_SUSC + SS_SUSC

Genetic Model
AA_births = A_FREQ$^\wedge$2 * Total_People * BR
AA_FREQ = AA_Total/Total_People
AS_births = 2 * A_FREQ * S_FREQ * Total_People * BR
AS_FREQ = AS_Total/Total_People
A_alleles = AA_Total * 2 + AS_Total
A_FREQ = A_alleles/Total_Alleles
BR = 0.03969
SS_births = S_FREQ$^\wedge$2 * Total_People * BR
SS_FREQ = SS_Total/Total_People
S_alleles = SS_Total * 2 + AS_Total
S_FREQ = S_alleles/Total_Alleles
Total_Alleles = A_alleles + S_alleles
Total_People = AS_Total + SS_Total + AA_Total

Normal Blood (AA) Malaria Infection
AA_IMMUNE(t) = AA_IMMUNE(t − dt) + (AA_RECOV − AA_IMM_LOSS − AA_IM_D) * dt
INIT AA_IMMUNE = 1

INFLOWS:
AA_RECOV = AA_INF * RECOV_R
OUTFLOWS:
AA_IMM_LOSS = AA_IMMUNE * Immunity_Loss
AA_IM_D = AA_IMMUNE * Nat_DR
AA_INF(t) = AA_INF(t − dt) + (AA_INFECT − AA_RECOV − AA_I_D) * dt
INIT AA_INF = 36

INFLOWS:
AA_INFECT = AA_SUSC*Trans_AA * INF_LAG
OUTFLOWS:
AA_RECOV = AA_INF * RECOV_R
AA_I_D = AA_INF * INF_DR

AA_SUSC(t) = AA_SUSC(t − dt) + (AA_birth + AA_IMM_LOSS −
AA_INFECT − AA_S_D) * dt
INIT AA_SUSC = 94

INFLOWS:
AA_birth = AA_births
AA_IMM_LOSS = AA_IMMUNE * Immunity_Loss
OUTFLOWS:
AA_INFECT = AA_SUSC * Trans_AA * INF_LAG
AA_S_D = AA_SUSC * Nat_DR
Infected_Vector = 1 + SINWAVE(0.5,1) + RANDOM(-0.2, 0.2)
INF_DR = 0.1127
INF_LAG = DELAY(INFECTED, 0.0274)
Nat_DR = 0.01391
normal_TR = 0.001
Trans_AA = Infected_Vector * normal_TR
IMMUNITY_LOSS = GRAPH(Percent_Infected)
(0.00, 0.955), (0.167, 0.66), (0.333, 0.45), (0.5, 0.23), (0.667, 0.1), (0.833,
0.055), (1, 0.045)

Sickle Cell (SS) Malaria Infection
SS_IMMUNE(t) = SS_IMMUNE(t − dt) + (SS_RECOV − SS_IM_LOSS −
SS_IM_D) * dt
INIT SS_IMMUNE = 0

INFLOWS:
SS_RECOV = SS_INF*RECOV_R
OUTFLOWS:
SS_IM_LOSS = SS_IMMUNE*Immunity_Loss
SS_IM_D = SS_IMMUNE*Anemia_DR
SS_INF(t) = SS_INF(t − dt) + (SS_INFECT − SS_RECOV − SS_I_D) * dt
INIT SS_INF = 0

INFLOWS:
SS_INFECT = SS_SUSC * Trans_SS * INF_LAG
OUTFLOWS:
SS_RECOV = SS_INF * RECOV_R
SS_I_D = SS_INF * INF_DR+SS_INF * Anemia_DR
SS_SUSC(t) = SS_SUSC(t − dt) + (SS_birth + SS_IM_LOSS − SS_INFECT −
SS_S_D) * dt
INIT SS_SUSC = 1

INFLOWS:
SS_birth = SS_births
SS_IM_LOSS = SS_IMMUNE * Immunity_Loss
OUTFLOWS:
SS_INFECT = SS_SUSC * Trans_SS * INF_LAG

SS_S_D = SS_SUSC * Anemia_DR
Anemia_DR = 0.5
RECOV_R = 0.0125
Trans_SS = Infected_Vector * S_allele_TR

Chapter 5
Encephalitis

A virus transmitted between hosts, most commonly by mosquitoes and ticks, causes encephalitis. The disease is kept extant by a complicated transmission process between humans and a vertebrate host such as a horse. Humans and domestic animals are terminal hosts. They suffer from the disease but do not spread the disease themselves. Several paths of transmission and several variants of the virus are known. Most often, the transmitting vector (mosquito or tick, for example) will pass the virus to his or her own offspring.

Four distinct types of mosquito-transmitted encephalitis are found in the United States: western equine, eastern equine, St. Louis, and La Crosse. The time period for appearance of the disease is from June through September and even into winter in the warm parts of the country. The virus causes flu-like symptoms that are seldom fatal. Currently, antibiotics and vaccines approved by the Food and Drug Administration are ineffective. The disease is best prevented in humans by the use of specific pesticides and protective clothing and by avoiding the transmitting insects.

Detailed data for this serious public health problem have been collected in many parts of the country. Yet, relatively few models of the disease have been made. We offer one such model of St. Louis encephalitis in Illinois. Models of this disease are useful as public health instruments as they enable better prediction of the conditions for its appearance, and because they can be used to evaluate the effectiveness of intervention programs.

The combination of good models, an aware public, and an alert and efficient public health service can combine to effectively control this disease.

5.1 St. Louis Encephalitis

St. Louis encephalitis is the most common form of this disease in the United States, occurring in 48 states. Few infected people display symptoms, yet nearly 3,000 cases are reported annually. The older one is, the more likely one is to suffer or die from the infection, with an overall death rate of around 10 percent. Surveillance and

B. Hannon and M. Ruth, *Dynamic Modeling of Diseases and Pests*,
Modeling Dynamic Systems,
© Springer Science + Business Media LLC 2009

control of this disease costs about $150 million per year. The Centers for Disease Control predicts that global warming and the deterioration of the inner city could increase the prevalence of the disease[1].

Our model is based on St. Louis encephalitis in Illinois with much of the data coming from the work of Dr. Thomas Monath, adjunct of the Harvard School of Public Health[2]. The central guiding question is: How do different mosquito developmental thresholds affect the incidence of St. Louis encephalitis in the human population?

Generally, we had to model three populations—humans, birds, and mosquitoes— with the mosquito acting as the connection vector between the other two. Due to rapid changes in the mosquito populations, the emigration schedules of birds, and the rapidly changing temperatures, we chose a daily modeling time step with a Julian calendar for seasonal change, precipitation, and the timing of temperature changes. Temperature and available mosquito breeding sites (standing water) are the conditions that determine the reproduction success of the mosquito population. We created a simple model to produce a realistic but stochastic representation of temperature (Figure 5.1) and standing water (Figure 5.2).

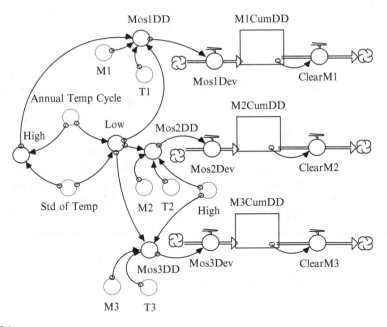

Fig. 5.1

---

[1] CDC website, Division of Vector-Borne Diseases, Domestic Arboviral Encephalitis, St. Louis Encephalitis, 13 July 2001.

[2] Monath, T.P. 1993. Arthropod-borne viruses. In: Morse, S.S., (Ed.). Emerging Viruses, Oxford University Press, New York, pp. 138–148. See also his volume by CRC on Arboviruses, 1989.

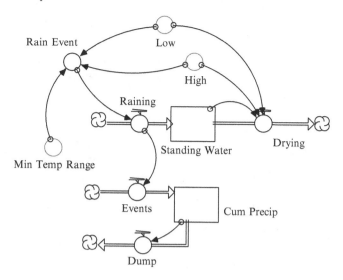

Fig. 5.2

The variable Annual Temp Cycle uses the STELLA built-in SINEWAVE function to reflect seasonality. This SINEWAVE function was adjusted to produce a maximum temperature on Julian Date 228 (Aug. 16) and a minimum on JD 45 (Feb. 14). The temperature range for the year is set at +/- 50 degrees F. The high and low temperatures, set independently from normal distributions, were: high = the average daily temperature + 1.5 * a standard deviation and low subtracting 1.5 * the same standard deviation from the mean. This method gave a variable but realistic temperature variation for Illinois.

Standing water is necessary for mosquito production. We include both rain events and evapotranspiration in the model as functions of the high and low temperature (Figure 5.2). We assume there is no probability of rain occurring when the difference between the high and low temperatures is greater than some threshold. We set that threshold at 8 degrees. If the daily temperature difference is below 8 degrees F, we choose a rain event from a normal distribution with an average of 0.8 inches of rain and a standard deviation of 0.4 inches. We find this result to be compatible with rain events in Illinois. Precipitation is accumulated to provide standing water, and standing water is removed from the system using a simple drying equation (Average Daily Temperature / 75 degrees) mimicking evaporation.

While the model allows for development of three species with different developmental thresholds, we implemented only one species (the third, indicated by the number 3 in the variable names) in the transmission cycle for this presentation with cool season development beginning at 50 degrees F. The corresponding module is shown in Figure 5.3.

Daily degree-day accumulations are used to schedule mosquito development when standing water is present. The mosquito population is divided into three age classes (eggs, larvae, and adults), and all have the same developmental temperature

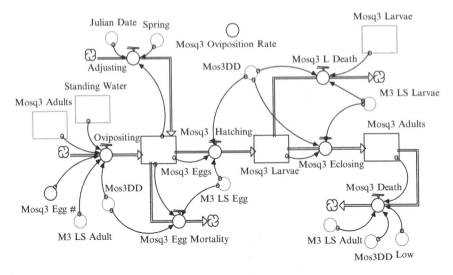

Fig. 5.3

threshold. Research on mosquito development[3] indicated that species differ in the proportion of their total life span spent in each age class. Some species spend over half their life as larvae while others have very short larval periods relative to their total life span (as measured in degree days). We believe this may be an important component of examining transmission with respect to the period of time when standing water is available in the environment. A short larval period allows for very quick response to the availability of breeding sites. Consequently, we chose to define mosquito development by the duration of the larval period as a proportion of the entire life span. We then assumed that egg duration is one-third the remaining life span and adult duration is two-thirds the remaining life span. The model can be refined, as more becomes know about the life span of particular mosquitoes.

Although there is evidence that arboviruses like St. Louis encephalitis and West Nile are present in other vertebrates (domestic horses, raccoons, and even reptiles), the titers detected in bird populations correspond best to the transmission of the disease to humans. Passeriforme (warblers, jays, crows, etc.) and Columbiforme (pigeons and doves) orders are specifically identified in the geographical spread and the amplification of the virus during the summer months[4]. We developed a simple seasonal reproduction bird model for susceptible birds assuming that most reproduction occurs on or around the spring equinox (Julian Date = 80). Avian reproduction appears to be timed to this annual date. Mortality begins immediately as an exponential decrease of the population at a rate that yields approximately the same bird population by Julian Date 80 of the following year (Figure 5.4). The populations and rates in this portion of the model are necessarily speculative and should

---

[3] See footnote 2.

[4] See footnote 2.

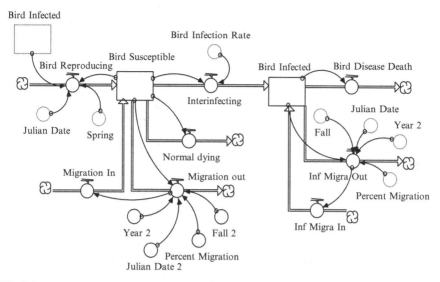

Fig. 5.4

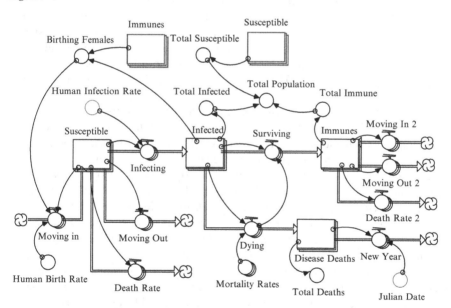

Fig. 5.5

be investigated in greater detail before drawing any practical conclusions from the model. We do know that the impact of the disease on the bird death rate is very small.

We divided the human population into cohorts by the degree of immunity and sickness and different ages, represented by the arrays in the module of Figure 5.5.

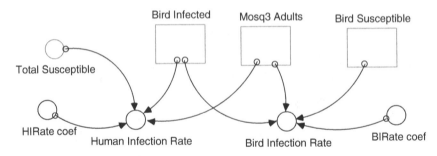

Fig. 5.6

This module shows the human population according to their susceptibility to encephalitis, the number with the disease at any given time, and the number of immunes, with each of these elements of the population inflows of people either through birth (Susceptibles) or inflows from a previous category. We sized the population for Illinois using U.S. Bureau of Census data for 2000.

Two infection rates are calculated (Figure 5.6). A bird infection rate is calculated as the product of the population of infectious birds, mosquito population, susceptible bird population, and coefficient of infection (adjusted to achieve transmission). Similarly, a human infection rate is calculated as the product of susceptible humans, mosquito population, infectious birds, and a coefficient of infection (adjusted to achieve human transmission). The variable Total Susceptible is the sum of the array of Human Susceptibles.

Several complications arise when matching the virus transmission rates to data on human infection and mortality. Monath[5] indicates that the ratio of asymptomatic to symptomatic infections may be as high as 338:1, and that transmission rates of the virus typically do not differ between the age cohorts of the human population even though the disease generally affects the elderly most. Actual mortality among symptomatic humans ranges from 0% (age 10–14) to 18% (age 75+). Mortality attributed to viral infection is highly correlated to the incidence of hypertension. Since cardiovascular disease increases with age, this may be the determining factor for mortality following viral infection. We did not model the incidence of hypertension in the population to determine the incidence of death. Instead, we chose to use the mortality rates for each human age cohort relative to symptomatic infection as reported by the Centers for Disease Control in Monath. Modeling the actual causes of mortality is far more complicated and will require considerably more research.

The annual known human infection rate for Illinois is also obtained from CDC data[6] (Table 5.1):

---

[5] See footnote 2.

[6] http://www.cdc.gov/ncidod/dvbid/arbor/arbocase.htm
Arboviral Encephalitis Cases Reported in Humans, by Type, United States, 1964-2000, CDC, Atlanta, GA.

Table 5.1

Year	Infected
1964	44
1965	1
1968	23
1974	6
1975	581
1976	14
1977	13
1980	5
1982	2
1983	2
1988	1
1993	2

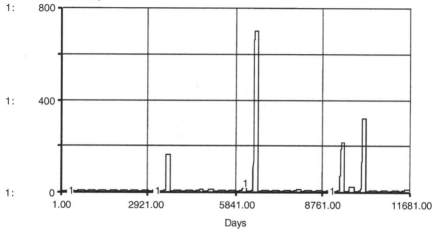

Fig. 5.7

In another report[7] an attack rate of 15.2 per 100,000 and case fatality of 9.7% is revealed. We adjusted the Monath human death rate factor (with a single coefficient) to get the death rate at about 10 percent of the death rate.

From Figure 5.7, we can see that the human infection rate was high in the first cycle of the two years of results even though the breeding potential for mosquitoes was equally high in both years. This difference in human impact is due to the large number of immunes in the population by the second summer of the model run (Figure 5.8). Note in Figure 5.8 that the bulk of the time for standing water does

---

[7] Hopkins, C.C., F.B. Hollinger, R.F. Johnson, H.J. Dewlett, V.F. Newhouse, and R.W. Chamberlain. 1975. The Epidemiology of St. Louis Encephalitis in Dallas, Texas, 1966. *American Journal of Epidemiology* 102: 1–15.

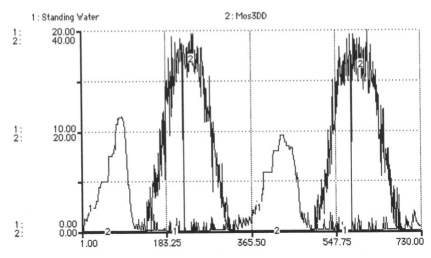

Fig. 5.8

not coincide with the peak in mosquito degree-days. Also note that the standing water peak does not coincide with the temperature peak in Illinois.

When running the model, we adjusted the HIRate Coef (Human Infection Rate Coefficient) and the BIRateCoef (Bird Infection Rate Coefficient) until we achieved an approximation of the infection rates over a 30-year period. While we could find 30-year periods where the model gave a reasonable approximation of the infection rate data, results from subsequent runs of the same model over the same period varied greatly, ranging from very few infections to a death rate exceeding 1,000. These results give us pause as to the effectiveness of the model and the possibility that the state has yet to experience its worst outbreak.

5.2 Questions and Tasks

1. One aspect of human behavior that is not included in the model is the fact that the state's Public Health Department has an effective warning/publicity policy. When an infection such as St. Louis encephalitis or West Nile virus begins early in the year, residents are continually warned about the dangers, and many modify their behavior to avoid mosquitoes. Include an early warning system in the model and re-run it. Interpret your results.
2. In the model created for this chapter, we assume that most mosquito-transmitted diseases will eventually become endemic in the population of birds and humans. However, we are interested in the short-term seasonal epidemics that may occur and what temporal control possibilities are feasible for prevention of epidemics in the human population. We think that this model is a good start in that direction.

Eventually, a more robust model would be able to address the potential effects of warmer, rainier climates and the extent of disease spread across the country.

a. How should such a model be set up?
b. Gradually change the climate in the model and observe the results.
c. Introduce more dramatic climatic changes (e.g. by including an increase in mean temperature, a decrease in precipitation, and increases in the standard deviation around the means.) How do your results differ from those obtained in part b?

3.a. Suppose that the typical annual confirmed case rate was 150 with 15 deaths. Adjust the model to approximately duplicate this result.
3b. Suppose global warming sets in. Illinois is expected to become drier and warmer. The mean temperature used in the annual cycle is 40 degrees F. Assume it is 50 degrees F due to global warming in 30 years. Let the mean raining event drop to .6 (see rainfall variable in the model) and run the calibrated model to assess the changes in human St. Louis encephalitis mortality due to global warming.

ENCEPHALITIS

ALL Controls
Annual_Temp_Cycle = (SIN((TIME+228)/58.1)*Annual_Temp_Range)+40
Annual_Temp_Range = 50
Fall = 264
Julian_Date = MOD(TIME, 365)
M1 = 90
M2 = 80
M3 = 90
M3_as_Larvae = 0.8
M3_LS_Adult = Mosq3_Life_Span * (1 − M3_as_Larvae) * 2 / 3
M3_LS_Egg = Mosq3_Life_Span * (1 − M3_as_Larvae) / 3
M3_LS_Larvae = M3_as_Larvae*Mosq3_Life_Span
Min_Temp_Range = 8
Mosq3_Life_Span = 499
Spring = 80
Std_of_Temp = 6
T1 = 65
T2 = 60
T3 = 50
Year = INT(TIME/365) + 1

Bird Population Dynamics
Bird_Infected(t) = Bird_Infected(t − dt) + (Bird_Infection + Inf_Migra_In −
Bird_Disease_Death − Inf_Migra_Out) * dt
INIT Bird_Infected = 0.001

INFLOWS:

Bird_Infection = if time > 0 then Bird_Suceptible*Bird_Infection_Rate else 0

Inf_Migra_In = DELAY(Inf_Migra_Out,183)

OUTFLOWS:

Bird_Disease_Death = if time > 0 then Bird_Infected*0.002 else 0

Inf_Migra_Out = If Year_2=Year_2 then if Julian_Date = Fall then Bird_Infected * Percent_Migration else 0 else 0

Bird_Suceptible(t) = Bird_Suceptible(t − dt) + (Bird_Reprod + Migration_In − Bird_Infection − Regular_Death − Migration_out) * dt

INIT Bird_Suceptible = 100

INFLOWS:

Bird_Reprod = If Julian_Date > 0 then IF Julian_Date = (Spring + 1) then 3 * (Bird_Infected+Bird_Suceptible) else 0 else 0

Migration_In = DELAY(Migration_out,183)

OUTFLOWS:

Bird_Infection = if time > 0 then Bird_Suceptible*Bird_Infection_Rate else 0

Regular_Death = If time > 0 then Bird_Suceptible*0.001 else 0

Migration_out = If Year_2=Year_2 then if Julian_Date_2 = Fall_2 then Bird_Suceptible * Percent_Migration else 0 else 0

Fall_2 = 264 {Fall Equinox − Sept 21st}

Julian_Date_2 = if time < 0 then 0 else MOD(TIME,365)

Percent_Migration = .75

Year_2 = INT(TIME/365) + 1

Calculate Infection

BIRate_Fudge = 1 / (10 ^ 7)

Bird_Infection_Rate = Bird_Suceptible * Bird_Infected * Mosq3_Adults * BIRate_Fudge

HIRate_Fudge = 1 / (10 ^ 13)

Human_Infection_Rate = Total_Susceptible * Mosq3_Adults * Bird_Infected * HIRate_Fudge

Human Population

Disease_Deaths[Dim_Name_1_1](t) = Disease_Deaths[Dim_Name_1_1](t − dt) + (Dying[Dim_Name_1_1] − New_Year[Dim_Name_1_1]) * dt

INIT Disease_Deaths[Dim_Name_1_1] = 0

INFLOWS:

Dying[Age_0_to_4] = Infected[Age_0_to_4]*Mortality_Rates[Age_0_to_4]/365

Dying[Age_5_to_9] = Infected[Age_5_to_9]*Mortality_Rates[Age_5_to_9]/365

Dying[Age_10_to_14] = Infected[Age_10_to_14]*Mortality_Rates[Age_10_to_14]/365

Dying[Age_15_to_24] = Infected[Age_15_to_24]*Mortality_Rates[Age_15_to_24]/365

Dying[Age_25_to_34] = Infected[Age_25_to_34]*Mortality_Rates[Age_25_to_34]/365

Dying[Age_35_to_44] =
Infected[Age_35_to_44]*Mortality_Rates[Age_35_to_44]/365
Dying[Age_45_to_54] =
Infected[Age_45_to_54]*Mortality_Rates[Age_45_to_54]/365
Dying[Age_55_to_64] =
Infected[Age_55_to_64]*Mortality_Rates[Age_55_to_64]/365
Dying[Age_65_to_74] =
Infected[Age_65_to_74]*Mortality_Rates[Age_65_to_74]/365
Dying[Age_75_plus] =
Infected[Age_75_plus]*Mortality_Rates[Age_75_plus]/365
OUTFLOWS:
New_Year[Dim_Name_1_1] = IF Julian_Date = 0 then
ARRAYSUM(Disease_Deaths[*]) else 0
Immunes[Dim_Name_1_1](t) = Immunes[Dim_Name_1_1](t − dt) +
(Surviving[Dim_Name_1_1] + Moving_In_2[Dim_Name_1_1] −
Moving_Out_2[Dim_Name_1_1] − Death_Rate_2[Dim_Name_1_1]) * dt
INIT Immunes[Dim_Name_1_1] = 0

INFLOWS:
Surviving[Age_0_to_4] = DELAY(Infected[Age_0_to_4] − Dying[Age_0_to_4],5)
Surviving[Age_5_to_9] = DELAY(Infected[Age_5_to_9] − Dying[Age_5_to_9],5)
Surviving[Age_10_to_14] =
DELAY(Infected[Age_10_to_14] − Dying[Age_10_to_14],5)
Surviving[Age_15_to_24] =
DELAY(Infected[Age_15_to_24] − Dying[Age_15_to_24],5)
Surviving[Age_25_to_34] =
DELAY(Infected[Age_25_to_34] − Dying[Age_25_to_34],5)
Surviving[Age_35_to_44] =
DELAY(Infected[Age_35_to_44] − Dying[Age_35_to_44],5)
Surviving[Age_45_to_54] =
DELAY(Infected[Age_45_to_54] − Dying[Age_45_to_54],5)
Surviving[Age_55_to_64] =
DELAY(Infected[Age_55_to_64] − Dying[Age_55_to_64],5)
Surviving[Age_65_to_74] =
DELAY(Infected[Age_65_to_74] − Dying[Age_65_to_74],5)
Surviving[Age_75_plus] =
DELAY(Infected[Age_75_plus] − Dying[Age_75_plus],5)
Moving_In_2[Age_0_to_4] = Immunes[Age_0_to_4]*0
Moving_In_2[Age_5_to_9] = Immunes[Age_0_to_4]*0.2/365
Moving_In_2[Age_10_to_14] = Immunes[Age_5_to_9]*0.2/365
Moving_In_2[Age_15_to_24] = Immunes[Age_10_to_14]*0.1/365
Moving_In_2[Age_25_to_34] = Immunes[Age_15_to_24]*0.1/365
Moving_In_2[Age_35_to_44] = Immunes[Age_25_to_34]*0.1/365
Moving_In_2[Age_45_to_54] = Immunes[Age_35_to_44]*0.1/365
Moving_In_2[Age_55_to_64] = Immunes[Age_45_to_54]*0.1/365

Moving_In_2[Age_65_to_74] = Immunes[Age_55_to_64]*0.1/365
Moving_In_2[Age_75_plus] = Immunes[Age_65_to_74]*0.1/365

OUTFLOWS:
Moving_Out_2[Age_0_to_4] = Immunes[Age_0_to_4]*0.2/365
Moving_Out_2[Age_5_to_9] = Immunes[Age_5_to_9]*0.2/365
Moving_Out_2[Age_10_to_14] = Immunes[Age_10_to_14]*0.2/365
Moving_Out_2[Age_15_to_24] = Immunes[Age_15_to_24]*0.1/365
Moving_Out_2[Age_25_to_34] = Immunes[Age_25_to_34]*0.1/365
Moving_Out_2[Age_35_to_44] = Immunes[Age_35_to_44]*0.1/365
Moving_Out_2[Age_45_to_54] = Immunes[Age_45_to_54]*0.1/365
Moving_Out_2[Age_55_to_64] = Immunes[Age_55_to_64]*0.1/365
Moving_Out_2[Age_65_to_74] = Immunes[Age_65_to_74]*0.1/365
Moving_Out_2[Age_75_plus] = Immunes[Age_75_plus]*0
Death_Rate_2[Age_0_to_4] = Immunes[Age_0_to_4] * 0.0088/365
Death_Rate_2[Age_5_to_9] = Immunes[Age_5_to_9]*.0088/365
Death_Rate_2[Age_10_to_14] = Immunes[Age_10_to_14]*.0088/365
Death_Rate_2[Age_15_to_24] = Immunes[Age_15_to_24]*.0088/365
Death_Rate_2[Age_25_to_34] = Immunes[Age_25_to_34]*.0088/365
Death_Rate_2[Age_35_to_44] = Immunes[Age_35_to_44]*.0088/365
Death_Rate_2[Age_45_to_54] = Immunes[Age_45_to_54]*.0088/365
Death_Rate_2[Age_55_to_64] = Immunes[Age_55_to_64]*.0088/365
Death_Rate_2[Age_65_to_74] = Immunes[Age_65_to_74]*.0088/365
Death_Rate_2[Age_75_plus] = Immunes[Age_75_plus]*.0088/365
Infected[Dim_Name_1_1](t) = Infected[Dim_Name_1_1](t − dt) +
(Infecting[Dim_Name_1_1] − Surviving[Dim_Name_1_1] −
Dying[Dim_Name_1_1]) * dt
INIT Infected[Dim_Name_1_1] = 0

INFLOWS:
Infecting[Dim_Name_1_1] =
Human_Infection_Rate*Susceptible[Dim_Name_1_1]

OUTFLOWS:
Surviving[Age_0_to_4] = DELAY(Infected[Age_0_to_4] − Dying[Age_0_to_4],5)
Surviving[Age_5_to_9] = DELAY(Infected[Age_5_to_9] − Dying[Age_5_to_9],5)
Surviving[Age_10_to_14] =
DELAY(Infected[Age_10_to_14] − Dying[Age_10_to_14],5)
Surviving[Age_15_to_24] =
DELAY(Infected[Age_15_to_24] − Dying[Age_15_to_24],5)
Surviving[Age_25_to_34] =
DELAY(Infected[Age_25_to_34] − Dying[Age_25_to_34],5)
Surviving[Age_35_to_44] =
DELAY(Infected[Age_35_to_44] − Dying[Age_35_to_44],5)
Surviving[Age_45_to_54] =
DELAY(Infected[Age_45_to_54] − Dying[Age_45_to_54],5)

Surviving[Age_55_to_64] =
DELAY(Infected[Age_55_to_64] − Dying[Age_55_to_64],5)
Surviving[Age_65_to_74] =
DELAY(Infected[Age_65_to_74] − Dying[Age_65_to_74],5)
Surviving[Age_75_plus] =
DELAY(Infected[Age_75_plus] − Dying[Age_75_plus],5)
Dying[Age_0_to_4] =
Infected[Age_0_to_4]*Mortality_Rates[Age_0_to_4]/365
Dying[Age_5_to_9] =
Infected[Age_5_to_9]*Mortality_Rates[Age_5_to_9]/365
Dying[Age_10_to_14] =
Infected[Age_10_to_14]*Mortality_Rates[Age_10_to_14]/365
Dying[Age_15_to_24] =
Infected[Age_15_to_24]*Mortality_Rates[Age_15_to_24]/365
Dying[Age_25_to_34] =
Infected[Age_25_to_34]*Mortality_Rates[Age_25_to_34]/365
Dying[Age_35_to_44] =
Infected[Age_35_to_44]*Mortality_Rates[Age_35_to_44]/365
Dying[Age_45_to_54] =
Infected[Age_45_to_54]*Mortality_Rates[Age_45_to_54]/365
Dying[Age_55_to_64] =
Infected[Age_55_to_64]*Mortality_Rates[Age_55_to_64]/365
Dying[Age_65_to_74] =
Infected[Age_65_to_74]*Mortality_Rates[Age_65_to_74]/365
Dying[Age_75_plus] =
Infected[Age_75_plus]*Mortality_Rates[Age_75_plus]/365

Susceptible[Age_0_to_4](t) = Susceptible[Age_0_to_4](t − dt) +
(Moving_in[Age_0_to_4] − Infecting[Age_0_to_4] − Moving_Out[Age_0_to_4] −
Death_Rate[Age_0_to_4]) * dt
INIT Susceptible[Age_0_to_4] = 876549

Susceptible[Age_5_to_9](t) = Susceptible[Age_5_to_9](t − dt) +
(Moving_in[Age_5_to_9] − Infecting[Age_5_to_9] − Moving_Out[Age_5_to_9] −
Death_Rate[Age_5_to_9]) * dt
INIT Susceptible[Age_5_to_9] = 929858

Susceptible[Age_10_to_14](t) = Susceptible[Age_10_to_14](t − dt) +
(Moving_in[Age_10_to_14] − Infecting[Age_10_to_14] −
Moving_Out[Age_10_to_14] − Death_Rate[Age_10_to_14]) * dt
INIT Susceptible[Age_10_to_14] = 905097

Susceptible[Age_15_to_24](t) = Susceptible[Age_15_to_24](t − dt) +
(Moving_in[Age_15_to_24] − Infecting[Age_15_to_24] −
Moving_Out[Age_15_to_24] − Death_Rate[Age_15_to_24]) * dt
INIT Susceptible[Age_15_to_24] = 1744845

Susceptible[Age_25_to_34](t) = Susceptible[Age_25_to_34](t − dt) +
(Moving_in[Age_25_to_34] − Infecting[Age_25_to_34] −
Moving_Out[Age_25_to_34] − Death_Rate[Age_25_to_34]) * dt
INIT Susceptible[Age_25_to_34] = 1811674

Susceptible[Age_35_to_44](t) = Susceptible[Age_35_to_44](t − dt) +
(Moving_in[Age_35_to_44] − Infecting[Age_35_to_44] −
Moving_Out[Age_35_to_44] − Death_Rate[Age_35_to_44]) * dt
INIT Susceptible[Age_35_to_44] = 1983870

Susceptible[Age_45_to_54](t) = Susceptible[Age_45_to_54](t − dt) +
(Moving_in[Age_45_to_54] − Infecting[Age_45_to_54] −
Moving_Out[Age_45_to_54] − Death_Rate[Age_45_to_54]) * dt
INIT Susceptible[Age_45_to_54] = 1626742

Susceptible[Age_55_to_64](t) = Susceptible[Age_55_to_64](t − dt) +
(Moving_in[Age_55_to_64] − Infecting[Age_55_to_64] −
Moving_Out[Age_55_to_64] − Death_Rate[Age_55_to_64]) * dt
INIT Susceptible[Age_55_to_64] = 1040633

Susceptible[Age_65_to_74](t) = Susceptible[Age_65_to_74](t − dt) +
(Moving_in[Age_65_to_74] − Infecting[Age_65_to_74] −
Moving_Out[Age_65_to_74] − Death_Rate[Age_65_to_74]) * dt
INIT Susceptible[Age_65_to_74] = 772247

Susceptible[Age_75_plus](t) = Susceptible[Age_75_plus](t − dt) +
(Moving_in[Age_75_plus] − Infecting[Age_75_plus] −
Moving_Out[Age_75_plus] − Death_Rate[Age_75_plus]) * dt
INIT Susceptible[Age_75_plus] = 727778

INFLOWS:
Moving_in[Age_0_to_4] = (Birthing_Females + Susceptible[Age_15_to_24] +
Susceptible[Age_25_to_34] + Susceptible[Age_35_to_44]) *
(Human_Birth_Rate/365)
Moving_in[Age_5_to_9] = Susceptible[Age_0_to_4]*0.2/365 +
(Birthing_Females * Human_Birth_Rate * 0)
Moving_in[Age_10_to_14] = Susceptible[Age_5_to_9]*0.2/365+
(Birthing_Females * Human_Birth_Rate * 0)
Moving_in[Age_15_to_24] = Susceptible[Age_10_to_14]*0.2/365 +
(Birthing_Females * Human_Birth_Rate * 0)
Moving_in[Age_25_to_34] = Susceptible[Age_15_to_24]*0.1/365 +
(Birthing_Females * Human_Birth_Rate * 0)
Moving_in[Age_35_to_44] = Susceptible[Age_25_to_34]*0.1/365 +
(Birthing_Females * Human_Birth_Rate * 0)
Moving_in[Age_45_to_54] = Susceptible[Age_35_to_44]*0.1/365 +
(Birthing_Females * Human_Birth_Rate * 0)
Moving_in[Age_55_to_64] = Susceptible[Age_45_to_54]*0.1/365+
(Birthing_Females * Human_Birth_Rate * 0)

Moving_in[Age_65_to_74] = Susceptible[Age_55_to_64]*0.1/365 +
(Birthing_Females * Human_Birth_Rate * 0)
Moving_in[Age_75_plus] = Susceptible[Age_65_to_74]*0.1/365 +
(Birthing_Females * Human_Birth_Rate * 0)

OUTFLOWS:
Infecting[Dim_Name_1_1] =
Human_Infection_Rate*Susceptible[Dim_Name_1_1]
Moving_Out[Age_0_to_4] = Susceptible[Age_0_to_4]*0.2/365
Moving_Out[Age_5_to_9] = Susceptible[Age_5_to_9]*0.2/365
Moving_Out[Age_10_to_14] = Susceptible[Age_10_to_14]*0.2/365
Moving_Out[Age_15_to_24] = Susceptible[Age_15_to_24]*0.1/365
Moving_Out[Age_25_to_34] = Susceptible[Age_25_to_34]*0.1/365
Moving_Out[Age_35_to_44] = Susceptible[Age_35_to_44]*0.1/365
Moving_Out[Age_45_to_54] = Susceptible[Age_45_to_54]*0.1/365
Moving_Out[Age_55_to_64] = Susceptible[Age_55_to_64]*0.1/365
Moving_Out[Age_65_to_74] = Susceptible[Age_65_to_74]*0.1/365
Moving_Out[Age_75_plus] = Susceptible[Age_75_plus]*0
Death_Rate[Dim_Name_1_1] = Susceptible[Dim_Name_1_1]*.0088/365
Birthing_Females = (Infected[Age_15_to_24]+Infected[Age_25_to_34]+
Infected[Age_35_to_44])+(Immunes[Age_15_to_24]+
Immunes[Age_25_to_34]+Immunes[Age_35_to_44])/2
Human_Birth_Rate = 0.068
Mortality_Rates[Age_0_to_4] = 0.0256
Mortality_Rates[Age_5_to_9] = 0.0169
Mortality_Rates[Age_10_to_14] = 0
Mortality_Rates[Age_15_to_24] = 0.0148
Mortality_Rates[Age_25_to_34] = 0.0056
Mortality_Rates[Age_35_to_44] = 0.0179
Mortality_Rates[Age_45_to_54] = 0.0596
Mortality_Rates[Age_55_to_64] = 0.0674
Mortality_Rates[Age_65_to_74] = 0.0955
Mortality_Rates[Age_75_plus] = 0.1803
Total_Deaths = ARRAYSUM(Disease_Deaths[*])
Total_Immune = ARRAYSUM(Immunes[*])
Total_Infected = ARRAYSUM(Infected[*])
Total_Population = Total_Immune+Total_Infected+Total_Susceptible
Total_Susceptible = ARRAYSUM(Susceptible[*])

Mosquito Population Dynamics
Mosq3_Adults(t) = Mosq3_Adults(t − dt) + (Mosq3_Eclosing − Mosq3_Death)
* dt
INIT Mosq3_Adults = 0

INFLOWS:
Mosq3_Eclosing = Mosq3_Larvae * (Mos3DD / M3_LS_Larvae)

OUTFLOWS:
Mosq3_Death = IF Low < 32 then 0.8 * Mosq3_Adults else Mosq3_Adults *
(Mos3DD/ M3_LS_Adult) * 0.5
Mosq3_Eggs(t) = Mosq3_Eggs(t − dt) + (Ovipositing + Fudge −
Mosq3_Hatching −
Mosq3_Egg_Mortality) * dt
INIT Mosq3_Eggs = 100

INFLOWS:
Ovipositing = Mosq3_Adults * Mosq3_Egg_# * (Mos3DD/ M3_LS_Adult) *
Standing_Water
Fudge = IF Julian_Date = Spring and Mosq3_Eggs < 100 then 100 −
Mosq3_Eggs else 0

OUTFLOWS:
Mosq3_Hatching = Mosq3_Eggs * (Mos3DD / M3_LS_Egg)
Mosq3_Egg_Mortality = if SMTH1(Mos3DD,30) = 0 then
(Mos3DD/M3_LS_Egg) * 0.1 * Mosq3_Eggs else (Mos3DD/M3_LS_Egg) *
0.3 * Mosq3_Eggs
Mosq3_Larvae(t) = Mosq3_Larvae(t − dt) + (Mosq3_Hatching −
Mosq3_Eclosing − Mosq_L_Death) * dt
INIT Mosq3_Larvae = 0

INFLOWS:
Mosq3_Hatching = Mosq3_Eggs * (Mos3DD / M3_LS_Egg)

OUTFLOWS:
Mosq3_Eclosing = Mosq3_Larvae * (Mos3DD / M3_LS_Larvae)
Mosq_L_Death = if SMTH1(Mos3DD,30)= 0 then 0.1 * Mosq3_Larvae else
(Mos3DD / M3_LS_Larvae) * 7 * Mosq3_Larvae
Mosq3_Egg_# = 100
Mosq3_Oviposition_Rate = 100

Precipitation
Cum_Precip(t) = Cum_Precip(t − dt) + (Events − Dump) * dt
INIT Cum_Precip = 0

INFLOWS:
Events = Raining

OUTFLOWS:
Dump = IF INT(TIME()/365)*365 = TIME then Cum_Precip else 0
Standing_Water(t) = Standing_Water(t − dt) + (Raining − Drying) * dt
INIT Standing_Water = 0

INFLOWS:
Raining = IF Rain_Event < 1 then Normal(0.8,0.4) else 0

OUTFLOWS:

Drying = Standing_Water*((High+Low)/2)/75
Rain_Event = (HIGH − LOW)/Min_Temp_Range
Temperature − Mosquito Development
M1CumDD(t) = M1CumDD(t − dt) + (Mos1Dev − ClearM1) * dt
INIT M1CumDD = 0

INFLOWS:
Mos1Dev = If Mos1DD<0 then 0 else Mos1DD
OUTFLOWS:
ClearM1 = If (INT(TIME/365)*365)=TIME then M1CumDD else 0
M2CumDD(t) = M2CumDD(t − dt) + (Mos2Dev − ClearM2) * dt
INIT M2CumDD = 0

INFLOWS:
Mos2Dev = If Mos2DD<0 then 0 else Mos2DD
OUTFLOWS:
ClearM2 = If (INT(TIME/365)*365)=TIME then M2CumDD else 0
M3CumDD(t) = M3CumDD(t − dt) + (Mos3Dev − ClearM3) * dt
INIT M3CumDD = 0

INFLOWS:
Mos3Dev = If Mos3DD<0 then 0 else Mos3DD
OUTFLOWS:
ClearM3 = If (INT(TIME/365)*365)=TIME then M3CumDD else 0
High = NORMAL(Annual_Temp_Cycle+1.5*Std_of_Temp,Std_of_Temp)
Low = NORMAL(Annual_Temp_Cycle−1.5*Std_of_Temp,Std_of_Temp)
Mos1DD = IF (HIGH < T1 OR LOW > M1) then 0 else
IF (HIGH < M1 AND LOW > T1) then (High+Low)/2 − T1 else
IF (HIGH > M1 AND LOW > T1) then (M1+Low)/2 − T1 else
IF (HIGH < M1 AND LOW < T1) then (HIGH+T1)/2 − T1 else 0
Mos2DD = IF (HIGH < T2 OR LOW > M2) then 0 else
IF (HIGH < M2 AND LOW > T2) then (High+Low)/2 − T2 else
IF (HIGH > M2 AND LOW > T2) then (M2+Low)/2 − T2 else
IF (HIGH < M2 AND LOW < T2) then (HIGH+T2)/2 − T2 else 0
Mos3DD = IF (HIGH < T3 OR LOW > M3) then 0 else
IF (HIGH < M3 AND LOW > T3) then (High+Low)/2 − T3 else
IF (HIGH > M3 AND LOW > T3) then (M3+Low)/2 − T3 else
IF (HIGH < M3 AND LOW < T3) then (HIGH+T3)/2 − T3 else 0

Chapter 6
Chagas Disease

Chagas disease, also called American Trypanosomiasis, is one of the most prominent vector-borne diseases in Latin America. It is caused by the protozoan parasite, *Trypanosoma cruzi*, and transmitted by blood-feeding triatomine bugs. Infection occurs not by the biting of the bugs, but by infiltration of the feces of infected bugs during blood feeding.

In rural areas of Argentina, for example, most people acquire infection in their homes because the cracks in the walls of their houses and the thatched roofs under which people live tend to be shelters for triatomine bugs. Furthermore, domestic and peridomestic animals, such as dogs and chickens, often occupy people's bedrooms, thus providing sufficient blood sources for the bugs. Human-to-human transmission is generally impossible, but a few infants are infected by their mothers, who already have Chagas disease, and in rare cases people can contract Chagas disease through blood transfusions. Because of the central role of animal-to-human transmission, more focus on *Trypanosomiasis infestans* as a main vector of the Chagas disease in domestic housing of Argentina[1] is expected, even though many species of triatomine bugs transmit *Trypanosomiasis cruzi* throughout the geographic region.

Currently, neither vaccine nor prophylaxis is available. An effective drug treatment is used for humans only in the acute and early chronic phase of infection. People cannot develop immunity for Chagas disease even after recovering from the infection.

Possible control measures for Chagas disease are 1) improving housing, 2) annual spraying of insecticides and killing of triatomine bugs, 3) providing early prevention for infants born to infected mothers through congenital surveillance and treatment of infected infants, 4) excluding reservoir animals from human housing, and 5) blood screening.

---

[1] Cecere, M.C., Castanera, M.B., Canale, D.M., Chuit, R. and E. Gurtler. 1999. *Trypanosoma cruzi* infection in Triatoma infestans and other triatomines: long-term effects of a control program in rural northwestern Argentina. Pan Am J Public Health, 5(6), 392–399.

B. Hannon and M. Ruth, *Dynamic Modeling of Diseases and Pests*,
Modeling Dynamic Systems,
© Springer Science + Business Media LLC 2009

6.1 Chagas Disease Spread and Control Strategies

The model of this chapter can be used to explore how Chagas disease spreads in a closed human population and how pursuing one or more of the following control strategies might affect spread of the disease. These control strategies include the following: improved housing structures, annual spraying, reservoir control, and congenital surveillance. The model consists of two sectors, one for the human population showing the transmission of *T. cruzi* among humans. The second module shows the life cycle of vector insects, and the transmission of *T. cruzi* from adults to infected vectors.

The human population module (Figure 6.1) is based on 2002 estimated Argentina's population size[2], BIRTHRATE, and OVERALL DEATH RATE (NATURAL DEATH RATE). We estimated the REPRODUCTIVE POPULATION from the total population and also estimate the age and sex distribution of the population. The initial number of SUSCEPTIBLE and INFECTED HUMANS is calculated from the total population and prevalence rate of *T. cruzi* in Argentina[3].

A human may acquire infection if borne to an infected mother. CONGENITAL INFECTION RATE is calculated as the product of PREVALENCE OF *T. CRUZI* AMONG PREGNANT MOTHERS (0.055) and CONGENITAL TRANSMISSION RATE (0.067)[4]. The majority of infected persons become infected by infected triatomine bugs. The HUMAN INFECTION RATE is the probability that susceptible humans become infected from these infection vectors:

$$\text{HUMAN INFECTION RATE} = 1 - (1 - \text{TRBH})^{\wedge}(\text{CONT INF B}), \qquad (6.1)$$

where TRBH is the probability that feeding of a bug on an uninfected human will cause the human to acquire infection at a rate of 0.0008[5].

CONT INF B is the average number of times an infected bug has had feeding contact with a human. It is the product of the NUMBER OF INFECTED VECTORS and the AVERAGE NUMBER OF FEEDINGS PER BUG EACH YEAR (5 times per year) divided by (TOTAL BLOOD SOURCE).[5] INFECTED HUMAN is obtained from the product of HUMAN INFECTION RATE and SUSCEPTIBLE HUMAN.

---

[2] http://www.cia.gov/cia/publications/factbook/geos/ar.html

[3] Gurtler, R.E., Cecere, M.C., Castanera, M.B., Canale, D., Lauricella, M.A., Chuit R., Cohen, J.E. and E.L. Segura. 1996. Probability of infection with *Trypanosoma cruzi* of the vector Triatoma infestans fed on infected humans and dogs in northwest Argentina. Am J Trop Med Hyg, 55(1), 24–31.

[4] Blanco, S.B., Segura, E.L., Cura, E.L., Chuit, R., Tulian, L., Flores, I., Garbarino, G., Villalonga, J.F. and R.E. Gurtler. 2000. Congenital transmission of *Trypanosoma cruzi*: an operational outline for detecting and treating infected infants in northwestern Argentina. Trop Med Int Health, 5(4), 293–301.

[5] Cohen, J.E. and R.E. Gurtler. 2001. Modeling household transmission of American Trypanosomiasis. Science, 293, 694–698.

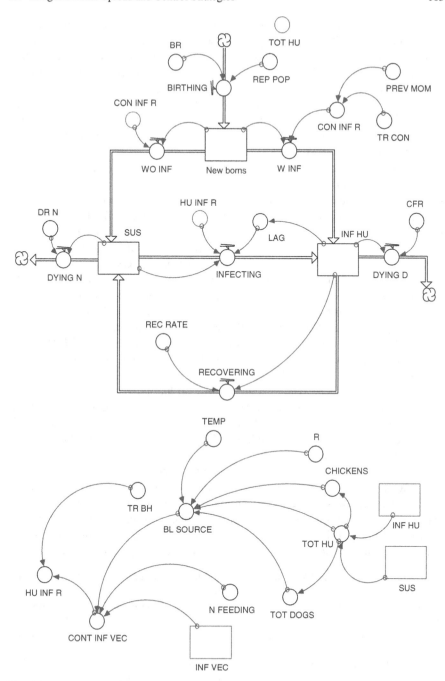

Fig. 6.1

Infected people become infectious to others after a 10-day (0.03) lag time[6] Of those infected, 14–44% die of Chagas disease. This range of case fatality depends on the person's age, when infection occurs, and the stage of infection. In this model, however, we treated fatality as a random variable, in part because of a lack of data on disease incidence by the various age cohorts of the population.[7] Approximately 60% of infected people recover once treated, although the recovery rate of the disease varies according to age of infected person and progression of the disease[8]. Additionally, we assume that there is no human infection via transfusion of contaminated blood.

The vector population module shows the life cycle of vector insects, and the transmission of *T. cruzi* from adults to infected vectors (Figure 6.2). The initial number of adult insects was estimated from the total human population and household bug density based on a field study (1,299 bugs/34 household)[9]. This household bug density was obtained on the assumption that these households had poor housing condition with no previous insecticide exposure. We also assumed that only uninfected bugs exist at the initial time step. Other parameters used to explain the life cycle of *T. cruzi* infesting bugs were based on laboratory experiments[10] as follows: The maximum number of uninfected and infected adult bugs that the physical infrastructure of each household will support, given an unlimited food supply (the carrying capacity) is K=500.[5] The monthly egg reproduction rate of each bug (EGG LAY RATE) is 1.64 and the egg survival fraction (ESF) is 0.57. NYMPH SURVIVAL RATE and ADULT SURVIVAL RATE are 0.175 and 307.5/365 per year, respectively.

Environmental effects, such as temperature and humidity, upon the survival of insects at each stage were not considered since the human housing provides a stable environment for bugs. Also, it was assumed that all bugs mature to the next stage, unless they die. Adult bugs become infected with *T. cruzi* through feeding contact with infected blood sources.

T. cruzi infestans are not mobile insects. Once sheltered at domiciliary areas, these bugs are not likely to move outside of the area to feed themselves. Thus, we assumed that bugs feed on vertebrate animals found in domiciliary areas. Humans and family dogs that are easily accessible to human housing are all-season blood sources for triatomine bugs, while chickens are blood sources only during relatively cold seasons. In addition, bugs feed on dogs and chickens approximately three times

---

[6] Markell E.K., John D.T. and W.A. Krotoski. 1999. Medical Parasitology. 8th Ed. Other Blood- and Tissue dwelling protozoa. (pp.134–146). PA: WB Saunders.

[7] Guhl, F. and G.A. Vallejo. 1999. Interruption of Chagas disease transmission in the Andean Countries: Colombia. Mem Inst Oswaldo Cruz, Rio de Janeiro, 94(supple.1), 413–415.

[8] Cancado, J.R. 1999. Criteria of Chagas disease cure. Mem Inst Oswaldo Cruz, Rio de Janeiro, 94(supple.1), 331–335.

[9] Gurtler, R.E., Cohen, J.E., Cecere, M.C., Lauricella, M.A., Chuit, R. and E.L. Segura. 1998. Influences of humans and domestic animals on the household prevalence of *Trypanosoma cruzi* and Triatoma infestans in northwest Argentina. Am J Trop Med Hyg, 58(6), 748–758.

[10] Guarneri, A.A., Carvalho Pinto, C.J., Schofield, C.J. and M. Steindel. 1923. Population biology of Rhodnius domesticus (Hemiptera: Reduviidae) under laboratory conditions. Departamento de Microbiologia e Parasitologia, Universidade Federal de Santa Catarina, 93(2), 273–276.

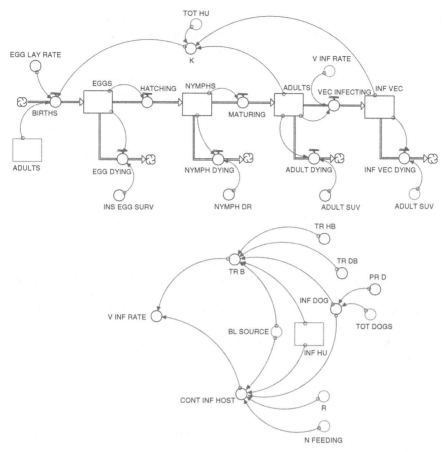

Fig. 6.2

as much as on humans, since it is not likely to be interrupted by these animals during feeding[11]. Therefore, we obtained BLOOD SOURCES of bugs as follows:

BLOOD SOURCE

= TOTAL HUMANS (SUSCEPTIBLE + INFECTED HUMANS)

+R ∗ (TOTAL DOGS + CHICKENS),

when TEMPERATURE < 70

and

BLOOD SOURCE = TOTAL HUMANS + R ∗ (TOTAL DOGS),

when TEMPERATURE <= 70, (6.2)

---

[11] Gurtler, R.E., Cohen, J.E., Cecere, M.C. and R. Chuit. 1997. Shifting host choices of the vector of Chagas disease, Triatoma infestans, in relation to the availability of hosts in houses in northwest Argentina. J Appl Ecology, 34, 699–715.

where R = 3 is the relative feeding index of dogs and chickens as sources of feeding contacts compared to humans and the temperature follows a sine wave with an annual average temperature of Argentina of 63 degrees F and a temperature range of 52 to 72 degrees.

The initial number of dogs and chickens are estimated from the total human population. We assumed that each household is comprised of 5 people, 2 dogs, and 2 chickens that are accessible to each human house.

Transmission of *T. cruzi* among insect vectors was determined as follows:

$$\text{VECTOR INFECTION RATE} = 1 - (1 - \text{TR B})^{\wedge}(\text{CONT INF HOST}), \quad (6.3)$$

where

$$\text{TR B} = ((\text{TR HB} * \text{INF HU}) + (\text{TR DB} * \text{INF DOG}))/\text{BLOOD SOURCE} \quad (6.4)$$

TR HB is the probability that, in one feeding by an initially uninfected bug on an infected (seropositive) human, the bug acquires infection. It is 0.03.[5] TR DB is the probability that, in one feeding by an initially uninfected bug on an infected (seropositive) human, the bug acquires infection. It is 0.49.[5] INF DOG is the number of infected dogs, calculated from the TOTAL DOGS and PREVALENCE OF *T. CRUZI* AMONG DOGS (0.84)[12].

CONT INF HOST is the average number of times an adult bug has had feeding contact with an infected blood source:

$$\text{CONT INF HOST} = \text{N FEEDING} * (\text{INF HU} + \text{R} * \text{INF DOG})/ \\ \text{BLOOD SOURCE} \quad (6.5)$$

However, chickens are not included in infected blood sources because they are not infected with *T. cruzi*.

The fatalities from *T. cruzi* among humans and the vector population dynamics are shown in Figures 6.3 and 6.4, respectively. When there is no control of *T. cruzi*, vector prevalence and the number of people dying of Chagas disease tend to increase through time.

When cracked and unplastered walls are plastered well, household bug density is decreased from 38 bugs to 0.43 bugs per household[13]. Since housing improvements are a one-time event, their introduction in the model changes the initial bug density and decreases the carrying capacity (K) from an assumed 500 to 200. This demonstrates the effects of housing improvements and is displayed in Figures 6.5 and 6.6. House improvement itself decreases the insect population and human

[12] Gurtler, R.E., Cecere, M.C., Rubel, D.N., Petersen, R.M., Schweigmann, N.J., Lauricella, M.A., Bujas, M.A., Segura, E.L. and C. Wisnivesky-Colli. 1991. Chagas disease in northwest Argentina: infected dogs as a risk factor for the domestic transmission of *Trypanosoma cruzi*. Transactions of the Royal Society of Tropical Medicine and Hygiene, 85, 741–745.

[13] Gurtler, R.E., Cecere, M.C., Rubel, D.N. and J. Schweigmann. 1992. Determinants of the domiciliary density of Triatoma infestans, vector of Chagas disease. Medical and Veterinary Entomology, 6, 75–83.

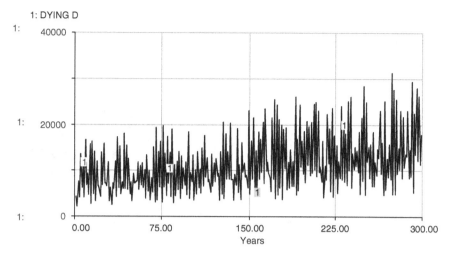

Fig. 6.3

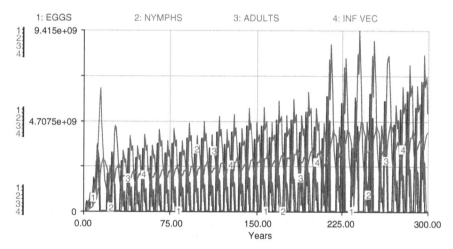

Fig. 6.4

infection rate and thus, decreases the number of deaths due to Chagas disease. Chagas disease cannot be eliminated solely by housing improvements, however.

The outflows from ADULT INSECT and INFECTED VECTOR were added to test the effect of annual insecticide spraying. Annual spraying was known to kill 93% of bugs in domiciliary areas.[12] The effect of annual spraying, together with improved housing, is displayed in Figures 6.7 and 6.8. The results show that regular insecticide use can eliminate the triatomine bugs from human housing as long as these bugs do not have the ability to develop resistance to insecticide.

Chickens and dogs are the two major blood sources besides humans. To see the effects of chickens and dogs on the spread of the disease, we decreased the number

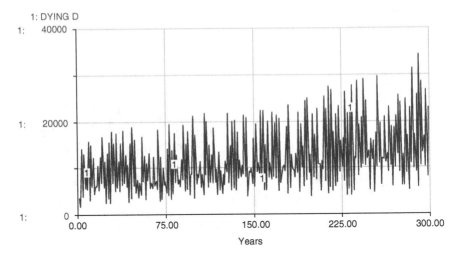

Fig. 6.5

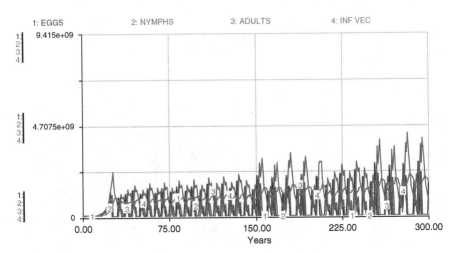

Fig. 6.6

of chickens per household to zero, and the number of dogs per household to one. Also, the Prevalence of *T. cruzi* among dogs was changed from 0.84 to 0.21 based on the assumption that *T. cruzi* prevalence among dogs decreased through dog surveillance programs.[13] The effects of reservoir controls are displayed Figure 6.9. The Figure 6.9 indicates that the exclusion of chickens and dogs means that bugs are left only to feed on humans; therefore, the infection of humans with *T. cruzi* increases.

The effectiveness of *T. cruzi* treatment varies according to age at which infection occurs and the stage of the infection. Early detection of congenitally infected infants ensures a recovery rate of over 90%.[4] The effects of congenital surveillance were tested by decreasing the CONGENITAL INFECTION RATE by 90%. This

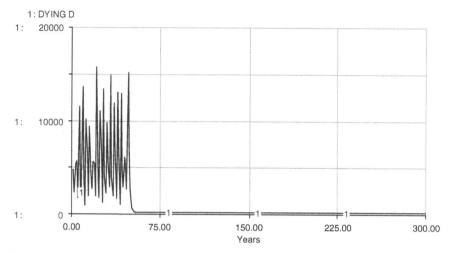

Fig. 6.7

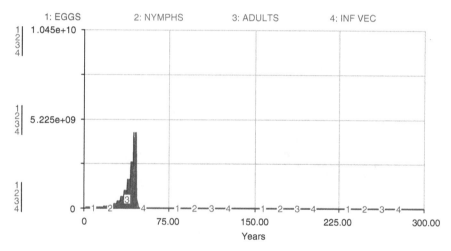

Fig. 6.8

assumes that congenital surveillance programs were successfully implemented, and the infected infants were treated with highly effective drugs. The effects of congenital surveillance are displayed in Figure 6.10. Congenital surveillance slightly decreases the HUMAN INFECTION RATE and the number of people dying of this disease. However, the impact is slight in terms of eliminating the disease.

Now that we have seen the impacts of different disease control strategies that are carried out individually, we can explore the effects of an aggressive control strategy that includes all of the control strategies previously detailed. The effects of the combined control strategy are displayed in Figure 6.11. The combined strategy causes almost the same effect on disease transmission as annual insecticide use. From this,

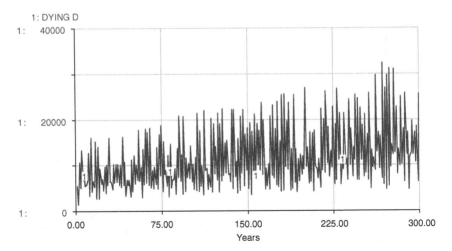

Fig. 6.9

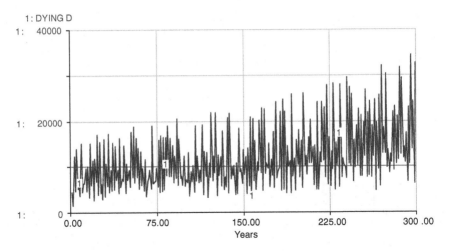

Fig. 6.10

we may conclude that annual insecticide use is the strongest measure of control that can be used to combat the transmission of Chagas disease in endemic areas.

6.2 Questions and Tasks

1. If triatomine bugs develop resistance to insecticide, how would the transmission of Chagas disease be affected?
2. Modify the model above to make different age cohorts of humans be affected by Chagas disease in different ways. For example, distinguish differences in recovery and mortality rates. What are the implications for transmission?

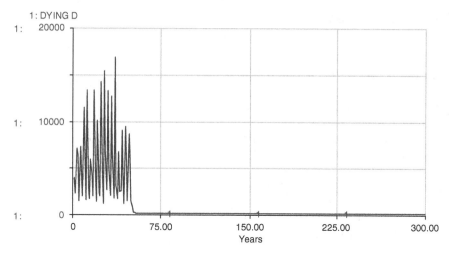

Fig. 6.11

3. How would the number of people living in the household affect the transmission of Chagas disease?
4. If temperature and relative humidity were factored into the model, would there be a significant effect in transmission?

CHAGAS DISEASE

HUMAN POPULATION
INF_HU(t) = INF_HU(t − dt) + (W_INF + INFECTING − RECOVERING − DYING_D) * dt
INIT INF_HU = 12931983

INFLOWS:
W_INF = New_borns*CON_INF_R
INFECTING = SUS*HU_INF_R*LAG
OUTFLOWS:
RECOVERING = REC_RATE*INF_HU
DYING_D = CFR*INF_HU
New_borns(t) = New_borns(t − dt) + (BIRTHING − W_INF − WO_INF) * dt
INIT New_borns = 0

INFLOWS:
BIRTHING = REP_POP*BR
OUTFLOWS:
W_INF = New_borns*CON_INF_R
WO_INF = New_borns*(1-CON_INF_R)
SUS(t) = SUS(t − dt) + (WO_INF + RECOVERING − INFECTING − DYING_N) * dt

INIT SUS = 24880834

INFLOWS:
WO_INF = New_borns*(1-CON_INF_R)
RECOVERING = REC_RATE*INF_HU
OUTFLOWS:
INFECTING = SUS*HU_INF_R*LAG
DYING_N = DR_N*SUS
BL_SOURCE = IF TEMP >= 70 THEN TOT_HU + R*TOT_DOGS ELSE
TOT_HU+R*(TOT_DOGS+CHICKENS)
BR = 18.23/1000
CFR = random(0.14/1000,0.44/1000)
CHICKENS = TOT_HU/5*2
CONT_INF_VEC = INF_VEC*N_FEEDING/BL_SOURCE
CON_INF_R = PREV_MOM*TR_CON
DR_N = 7.57/1000
HU_INF_R = 1 − (1 − TR_BH)∧CONT_INF_VEC
LAG = DELAY(INF_HU,0.03)
N_FEEDING = 5
PREV_MOM = 0.055
R = 3
REC_RATE = 0.6
REP_POP = TOT_HU*0.3*0.5
TEMP = SINWAVE(63,10)
TOT_DOGS = TOT_HU/5*2
TOT_HU = SUS + INF_HU
TR_BH = 0.0008
TR_CON = 0.067

VECTOR POPULATION
ADULTS(t) = ADULTS(t − dt) + (MATURING − ADULT_DYING −
VEC_INFECTING) * dt
INIT ADULTS = 288934408

INFLOWS:
MATURING = NYMPHS
OUTFLOWS:
ADULT_DYING = (1 − ADULT_SUV)*ADULTS
VEC_INFECTING = ADULTS*V_INF_RATE
EGGS(t) = EGGS(t − dt) + (BIRTHS − HATCHING − EGG_DYING) * dt
INIT EGGS = 0

INFLOWS:
BIRTHS = IF K<= 500 THEN EGG_LAY_RATE*ADULTS ELSE 0
OUTFLOWS:
HATCHING = EGGS
EGG_DYING = (1 − INS_EGG_SURV)*EGGS

INF_VEC(t) = INF_VEC(t − dt) + (VEC_INFECTING − INF_VEC_DYING) * dt
INIT INF_VEC = 0

INFLOWS:
VEC_INFECTING = ADULTS*V_INF_RATE
OUTFLOWS:
INF_VEC_DYING = (1 − ADULT_SUV)*INF_VEC
NYMPHS(t) = NYMPHS(t − dt) + (HATCHING − MATURING − NYMPH_DYING) * dt
INIT NYMPHS = 0

INFLOWS:
HATCHING = EGGS
OUTFLOWS:
MATURING = NYMPHS
NYMPH_DYING = NYMPH_DR*NYMPHS
ADULT_SUV = .84
CONT_INF_HOST = N_FEEDING*(INF_HU + R*INF_DOG)/BL_SOURCE
EGG_LAY_RATE = 1.64
INF_DOG = PR_D*TOT_DOGS
INS_EGG_SURV = .57
K = (ADULTS + INF_VEC)/(TOT_HU/5)
NYMPH_DR = .175
PR_D = 0.84
TR_B = ((TR_HB*INF_HU)+(TR_DB*INF_DOG))/BL_SOURCE
TR_DB = 0.49
TR_HB = 0.03
V_INF_RATE = 1 − (1 − TR_B)^CONT_INF_HOST

Chapter 7
Lyme Disease[1]

7.1 Lyme Disease Model

Over the past 20 years since Lyme disease was first diagnosed, it has been identified as the most common vector-borne disease in the United States. The repopula-
tion of white-tailed deer in the United States of America has been associated with
the emergence of this disease. The tick vector, *Ixodes scapularis,* harbors *Borre-
lia burgdorferi (B.b.),* the organism responsible for Lyme disease[2]. The larval and
nymphal stages feed on intermediate hosts, which are mostly small mammals and
birds. The adult tick prefers to feed on deer, but will also feed on dogs and people.
The main intermediate host in the northeast United States is the white-footed mouse.
Mice and chipmunks may serve as reservoirs for *B.b.* in nature since they maintain
active infections for at least 3 to 4 months. In the Midwest, however, it has been
determined that the eastern chipmunk may be equally important as an intermediate
host. The ticks appear to follow the migration of deer but deer may be simply an
amplification host; they are able to clear infection from *B.b.* within a few days.

To investigate the ecology of Lyme disease, we model a 1-hectare (0.1 km^2)
oak forest and run that model on a weekly time interval, beginning with the first
week of spring (T = 0), and DT = 0.5. In our model we concentrate on the interrelationships among acorn production, white-footed mice (*Peromyscus leucopus*),
white-tailed deer (*Odocoileus virginianus*), and the blacklegged tick (*Ixodes scapularis*). Specifically, we explore a hypothesis recently put forth by Ostfeld, Jones,
and Wolff,[3] which states that the risk for human Lyme disease, due to increases in
nymphal tick activity, should be greater approximately two years following a large
mast event.

---

[1] Thanks to M. Roberto Cortinas for helping develop the model of this chapter.

[2] Randolph, S.E. General Framework for Comparative Quantitative Studies on Transmission of
Tick-Borne Diseases Using Lyme Borreliosis in Europe as an Example. *J. Med. Entom.* 32(6)1995,
767–77.

[3] Ostfeld, Richard S., Clive G. Jones, and Jerry O. Wolff. 1996. Of mice and mast. *Bioscience* 46,
5: 323–330.

B. Hannon and M. Ruth, *Dynamic Modeling of Diseases and Pests,*
Modeling Dynamic Systems,
© Springer Science + Business Media LLC 2009

This hypothesis was developed based on the white-footed mouse population research performed by Wolff,[4] showing that summer mouse populations correlate with acorn production. Accordingly, if the summer mouse density increases following a large masting event, then larval ticks should not have any difficulty finding mice on which to obtain a blood meal—the authors seem to assume that the mouse density establishes the carrying capacity for larval ticks. Thus, many larvae will be able to complete molting into the nymphal stage and reappear the following spring (1.5 years after the masting event).

Unfed larvae generally do not carry the Lyme disease bacterial organism *B.d.* and so the active larvae season (summer into fall) is not correlated with human cases of Lyme disease. Larvae generally get the infection from mice that had been infected earlier in the year by the other generation's active nymphal stage (spring into summer). After winter stasis and early spring molting, the unfed nymph is ready for a blood meal and is likely to transmit the Lyme disease agent to an intermediate host or to a dead-end host, such as humans. Additionally, the unfed nymphal stage is the most dangerous, for it is the infected stage most likely to feed on human blood. Based on these facts, we should expect that if the population density of mice is high and many larvae can acquire a blood meal, the following spring should see a significant increase in active nymphs and an increase in human Lyme disease cases.

One key component of the ecology of Lyme disease is the production of acorns (Figure 7.1). We assume that acorns can be produced by two different species of oak trees, the white oak (*Quercus alba*) and the black oak (*Quercus velutina*). Based on the work by Sork et al.[5] in east-central Missouri, we have been able to simulate the average yearly acorn production for each tree species, as well as simulate the production of the large acorn mast event associated with the two species. The acorn production is normally distributed and nonnegative. The white oak has large acorn production about every three years, whereas the black oak has a large mast event every two years.

Based on the amount of total acorns produced, we have developed an assumption about ACORN MAST INDEX—an index that is ideally calculated on the basis of field observations by counting the number of acorns on a series of randomly selected branches on a series of randomly selected trees. However, we did not want to extrapolate how many branches a tree has. Instead, we assumed that if acorn production is zero, then the mast index is zero. Additionally, acorn mast indices above 200 seem to be rare, so we set the highest acorn production possible in our model to correlate with a mast index of 200. We assumed a linear relationship between acorn production and mast index. Because of this relationship and because the mast index is not a measure of a particular tree but a measure applied to several branches of several trees, we do not require a real value for the number of trees in our model, but a proportion of trees. Figure 7.2 shows an example model run for a 50% composition of both oak species in our forest patch.

---

[4] Wolff, Jerry O. 1996. Population fluctuations of mast-eating rodents are correlated with production of acorns. *Journal of Mammology* 77, 3: 850–856.

[5] Sork, Victoria L., Judy Bramble, and Owen Sexton. 1993. Ecology of mast-fruiting in three species of North American deciduous oaks. *Ecology* 74, 2: 528–541.

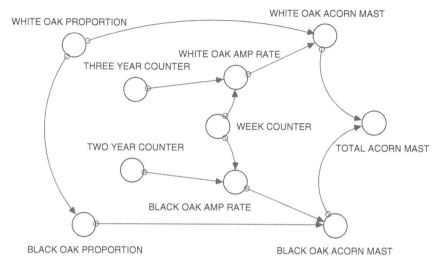

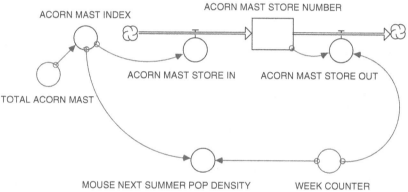

Fig. 7.1

Based on the work by Wolff, we used a linear regression equation that relates the summer white-footed mouse density (mice per hectare) to the acorn mast index:

$$\text{SUMMER MOUSE POPULATION DENSITY} =$$
$$7.28 + 0.60 * \text{ACORN MAST INDEX} \qquad (7.1)$$

Thus, the minimum mouse population density is 7.28, and mouse population density increases with an increase in the acorn mast index.

The mouse population dynamics are based on the acorn mast index (Figure 7.3). In order to use the above equation, we assumed the summer mouse population density to be the carrying capacity (K) for mice. Additionally, the reason for the increase in population (N) during the summer following a large mast event is due to an additional litter being born in the winter. White-footed mouse females generally have

1: ACORN MAST INDEX

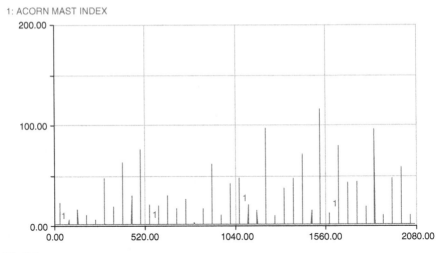

Fig. 7.2

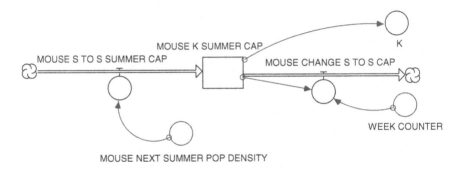

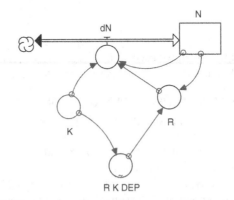

Fig. 7.3

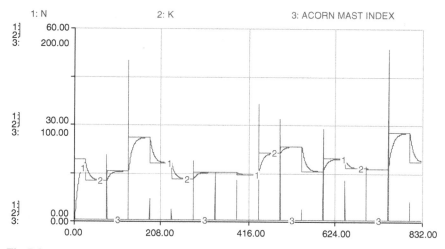

Fig. 7.4

two litters per year, but when the acorn mast is substantial some may have three. Because of this additional litter event, the natural rate of increase (R) for the population can vary based on the acorn mast production. Thus, we have assumed R to vary with changes in K. Since K is mast index-derived, R will also be driven by mast index due to its relationship with K. Nevertheless, the R value when K = 20 is based on Wolff's work.

Though we have made many assumptions that may not reflect the actual conditions in the mouse population, we have modeled a population that reaches the predicted density the summer following a mast event. Figure 7.4 shows the time-varying behavior of the mouse population size N, the carrying capacity, and the acorn mast index using Wolff's summer mouse population correlation equation. Figure 7.5 displays the relationship of R and the acorn mast index.

Changes in deer population dynamics are handled differently in the model from the ways in which we specified changes in mouse population (Figure 7.6). Based on the work by McShea and Schwede[6], we assumed deer respond to the variation in acorn crops by spending more or less time in particular parts of their habitats. The authors showed that deer are more likely to use the oak forest habitat if acorn production is good, and the percentage of habitat use is correlated with acorn production. Based on McShea and Schwede's data, we assumed an average 3.5 deer/ha. Their study area was approximately 60% nonforested and 40% forested. Thus, we could assume for the purpose of this model that if our model was placed in 1 km^2 of their study area, 4 ha would be forested and one of those hectares would be our model. Assume a deer had 50% habitat use in the forest. Since our model is 1/4 of the total forest, we assumed that the percentage of habitat used by deer in our model is 12.5% (1/2 * 1/4 = 1/8). We were able to roughly correlate the increases

---

[6] McShea, William J., and Georg Schwede. 1993. Variable acorn crops: Responses of white-tailed deer and other mast consumers. *Journal of Mammology* 74, 4: 999–1006.

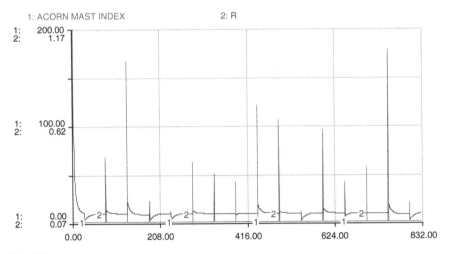

Fig. 7.5

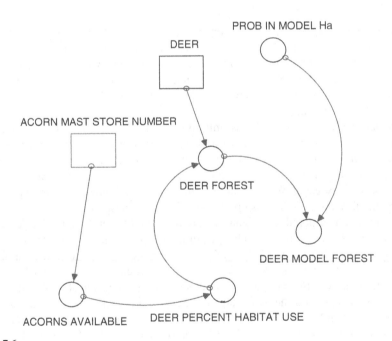

Fig. 7.6

in home range and forest habitat use with the acorn mast indices reported by Wolff. Both studies took place in Virginia at the same time, so it was good to see that when Wolff reported high acorn mast years, McShea and Schwede's data indicated increases in percent habitat use.

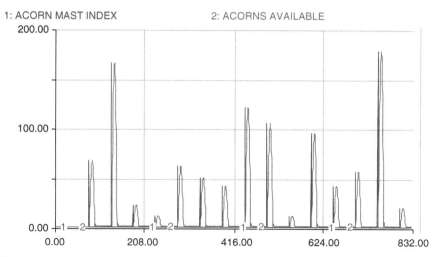

Fig. 7.7

Because DEER PERCENT HABITAT USE is a function of the presence of acorns in the forest, we modeled acorn fall by using a sine function (Figure 7.6). As soon as the acorn production is randomly selected in the oak-acorn sector (during the last week of summer) and the acorn mast index is generated, a sine wave of acorn availability develops, peaks during late September into October, and then declines. Since we roughly correlated the percent habitat use with the acorn mast index, we have assumed that acorn availability is proportional to the acorn mast index.

Based on the appearance and disappearance of acorns, the deer increase and then decrease their numbers in our model forest during the fall. Additionally, the maximum deer number is based on acorn production. In large mast years, more deer should come into our model. Figure 7.7 displays the acorn mast index in the late summer, and the subsequent autumnal acorn availability. Figure 7.8 shows the relationship between acorn mast index and deer wondering through our model.

Tick burdens on mice and deer are important, for we assumed that they represent the carrying capacity for our tick stage populations. Tick burdens on mice were assumed to be related to the number of tick larvae and nymphs found on mice during the spring and summer seasons. We used data collected from the 1990 and 1992–1997 seasons in Castle Rock State Park, Illinois. The data provided a mean number of larvae and nymphs on mice for a particular sampling day during a particular month. In order to ascertain what the tick burden would be for a particular week, we considered the biology of tick feeding. For example, larvae require about 4 days to successfully feed, whereas nymphs require about 4.5 days. In order to acquire the weekly larval tick burden for this part of the model (Figure 7.9) we specified the following relationship:

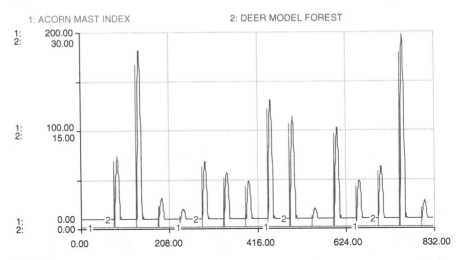

Fig. 7.8

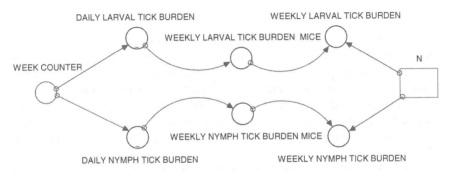

Fig. 7.9

$$\text{WEEKLY LARVAL TICK BURDEN} =$$
$$\text{N} * \text{DAILY LARVAL TICK BURDEN} * 7/4 \qquad (7.2)$$

For a daily larval tick burden of 5 larval ticks and 20 mice, 175 larval ticks could successfully feed per week (instead of 700 for a 1-day feeding time). The same mathematical assumption was applied to the weekly nymphal tick burden per mouse (Figure 7.9).

Figure 7.10 displays the relationship between mouse population density and the weekly larval and nymphal tick burdens. Note the seasonal distribution of nymphs and larvae. We used this seasonal information to regulate the appearance of unfed larvae and unfed nymphs. Although we omitted an environmental basis for the emergence of the nymphal and larval stages, including such factors as microclimate of the litter layer and humidity would make the model more realistic.

We assumed that the number of adult ticks entering and leaving our model was equal; consequently, we made no concessions to tick migration on deer.

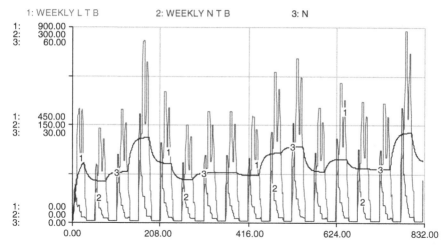

Fig. 7.10

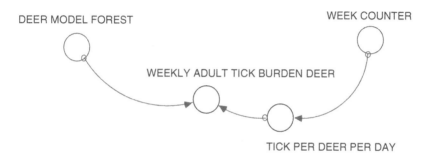

Fig. 7.11

Additionally, we assumed that deer were host to approximately 30 adult ticks per day. Since the feeding period for an adult female is 7 days, we assumed 30 adult ticks per week on a deer (Figure 7.11).

Figure 7.12 shows the seasonal variation in adult tick burdens in our 1-hectare forest.

The population dynamics of the blacklegged tick illustrate the two different populations that are active during particular seasons (Figures 7.13 and Figures 7.14). We assumed that the ticks cannot immigrate or emigrate from the model (or immigration = emigration). In nature, ovipositing and egg development are based on cumulative degree days. In lieu of good data on which to model temperature dependence, we approximated the development of eggs by employing a sine wave function. Average area under the curve is equal to the average success of hatching, and length (in weeks) is appropriate to the occurrence of hatching in the wild.

Ticks move in sequential fashion from egg to larva, to nymph, to adult during a 2-year period in which they are in resting stages prior to molting to the next

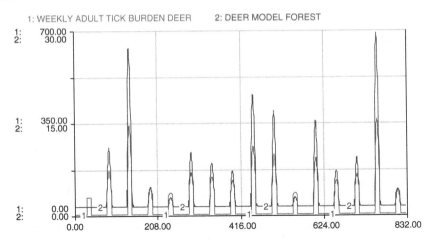

Fig. 7.12

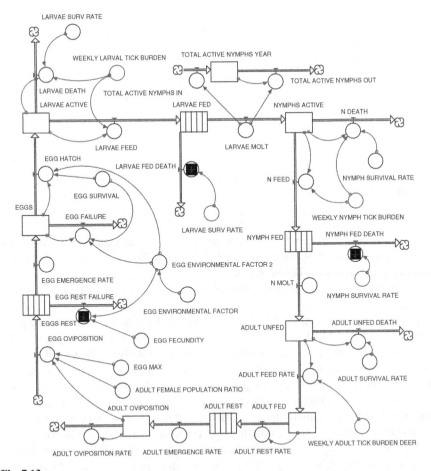

Fig. 7.13

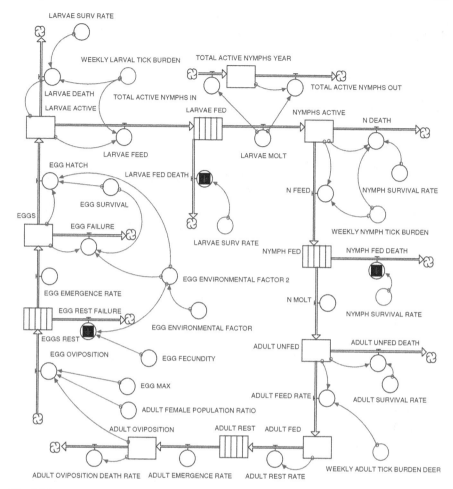

Fig. 7.14

life stage. Conveyors have been used to simulate the rest periods. Additionally, we assumed that intermediate and definitive host densities account for the carrying capacity and feeding success of the tick stages dependent on a blood meal to molt.

Figures 7.15–7.17 show, respectively, the populations of larvae, nymphs, and adult ticks. The latter indicates that the adult tick population numbers are not yet limited by deer numbers.

Now that all the pieces of the model are in place, we can calculate the number of active tick nymphs. By accumulating the number of ticks that are successfully molting from larval to nymphal form, we set up a reservoir that can provide the number of active nymphs per year. This variable can then be compared to the magnitude of the acorn mast index 2 years prior to see if the hypothesis holds true.

1: LARVAE ACTIVE   2: LARVAE ACTIVE 2   3: LARVAE FED   4: LARVAE FED 2   5: N

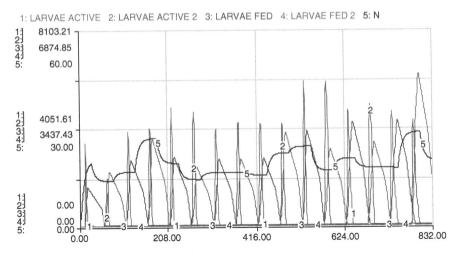

Fig. 7.15

1: NYMPHS ACTIVE   2: NYMPH ACTIVE 2   3: NYMPH FED   4: NYMPH FED 2

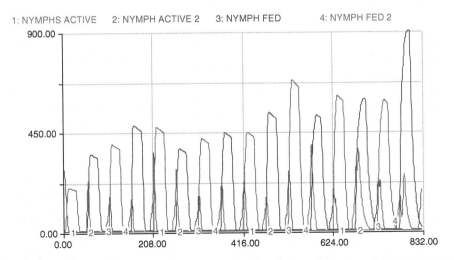

Fig. 7.16

Based on the results shown in Figure 7.18, it seems that nymphal activity increases almost 2 years following a large mast event. However, the increase is proportional to the difference between the mast event responsible for the increase and
the mast event that occurred almost 4 years prior to the nymphal increase. Thus, we
could make predictions about the danger of Lyme disease associated with a masting
event.

Another interesting question we can address with our model is whether the oak
species composition has a population effect on the two different tick populations.
I. scapularis has a lifespan of two years in the wild. If the populations are responsive
to acorn mast production, and excessive mast acorn production happens only every

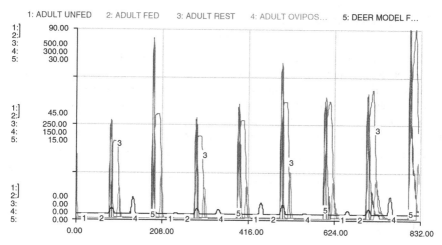

Fig. 7.17

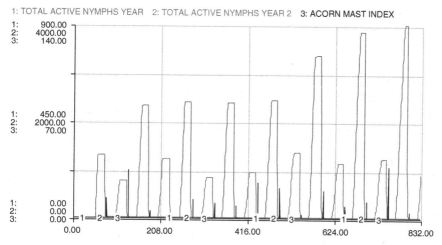

Fig. 7.18

two years, one population of ticks may crash or remain very low, whereas the other population maintains high numbers. As you run the model, you can see that population 1 benefits from the black oak mast, whereas population 2 maintains lower population numbers or crashes.

How about a combination? Because black oak masts are larger than white oak masts, a 50-50 species composition may still favor the tick population synchronous with the 2-year black oak mast. Even with white oak at 75%, population 2 crashes. At 100% white oak composition, you can see that the 3-year cycle of large masts benefits both tick populations. However, also note the beneficial effect on active nymphs not only 2 years after the mast event, but 3 years after the mast event.

7.2 Questions and Tasks

1. Could residual benefits occur from the large mast event that carried over because the tick burden carrying capacities were not limiting?
2. What would happen if mice and/or deer were excluded from the model forest?
3. What would happen if the deer population density changed in response to disease, predation, and harvesting?
4. What would happen if mice were allowed to emigrate / immigrate into the model?
5. The tick burdens on mice for the model of this chapter are based on Illinois data, and mouse tick burdens are higher on the East Coast. What if the model employed higher daily tick burdens on mice?

LYME DISEASE

Deer Movement
DEER(t) = DEER(t − dt)
INIT DEER = 35 {deer per 10 ha}

ACORNS_AVAILABLE =
ABS(SINWAVE(ACORN_MAST_STORE_NUMBER,(38 − 25)*2))*2
DEER_FOREST = DEER*DEER_PERCENT_HABITAT_USE/100
DEER_MODEL_FOREST = IF PROB_IN_MODEL_Ha < 0 THEN 0 ELSE
DEER_FOREST * PROB_IN_MODEL_Ha
PROB_IN_MODEL_Ha = 0.25 {Deer are in 10 ha = 1 sq. km, 6 ha nonforested,
4 ha forested. Model is in 1 of the 4 forested hectares, so there is a 1 in 4 (25%)
probability of being in the model forest.}
DEER_PERCENT_HABITAT_USE = GRAPH(ACORNS_AVAILABLE)
(0.00, 7.00), (10.0, 12.5), (20.0, 18.0), (30.0, 24.0), (40.0, 30.5), (50.0, 37.0),
(60.0, 44.0), (70.0, 50.5), (80.0, 57.5), (90.0, 64.0), (100, 70.5)

Mouse Population Dynamics

MOUSE_K_SUMMER_CAP(t) = MOUSE_K_SUMMER_CAP(t − dt) +
(MOUSE_S_TO_S_SUMMER_CAP − MOUSE_CHANGE_S_TO_S_CAP) * dt
INIT MOUSE_K_SUMMER_CAP = 20
INFLOWS:
MOUSE_S_TO_S_SUMMER_CAP =
MOUSE_NEXT_SUMMER_POP_DENSITY
OUTFLOWS:
MOUSE_CHANGE_S_TO_S_CAP = IF (TIME<25) THEN 0 ELSE IF
(WEEK_COUNTER=25) THEN MOUSE_K_SUMMER_CAP ELSE 0

N(t) = N(t − dt) + (dN) * dt
INIT N = 2
INFLOWS:
dN = R*N*(1-(N/K)) {Individuals per Week}

K = MOUSE_K_SUMMER_CAP
R = R_K_DEP/N
R_K_DEP = GRAPH(K)
(0.00, 0.00), (12.0, 1.40), (24.0, 2.80), (36.0, 3.50), (48.0, 4.20), (60.0, 4.90),
(72.0, 5.50), (84.0, 6.00), (96.0, 6.50), (108, 6.90), (120, 7.30)
Oak − Acorn Production
ACORN_MAST_STORE_NUMBER(t) =
ACORN_MAST_STORE_NUMBER(t − dt) +
(ACORN_MAST_STORE_IN − ACORN_MAST_STORE_OUT) * dt
INIT ACORN_MAST_STORE_NUMBER = 0
INFLOWS:
ACORN_MAST_STORE_IN = ACORN_MAST_INDEX
OUTFLOWS:
ACORN_MAST_STORE_OUT = IF WEEK_COUNTER>38 THEN
ACORN_MAST_STORE_NUMBER ELSE 0
BLACK_OAK_ACORN_MAST = If (BLACK_OAK_AMP_RATE < 0) THEN 0
ELSE BLACK_OAK_PROPORTION*BLACK_OAK_AMP_RATE
BLACK_OAK_AMP_RATE = IF (TWO_YEAR_COUNTER = 77) THEN
(NORMAL(3259,2983)) ELSE IF (TWO_YEAR_COUNTER<>77) AND
(WEEK_COUNTER=25) THEN (NORMAL(1059,834)) ELSE 0
BLACK_OAK_PROPORTION = IF WHITE_OAK_PROPORTION > 1
OR WHITE_OAK_PROPORTION < 0 THEN 1/0 ELSE
(1-WHITE_OAK_PROPORTION)
MOUSE_NEXT_SUMMER_POP_DENSITY = IF (WEEK_COUNTER = 25)
THEN (.60*ACORN_MAST_INDEX + 7.28) ELSE 0 {Number of mice the
following summer in one hectare}
THREE_YEAR_COUNTER = COUNTER(0,156)
TOTAL_ACORN_MAST =
BLACK_OAK_ACORN_MAST+WHITE_OAK_ACORN_MAST
TWO_YEAR_COUNTER = COUNTER(0,104)
WEEK_COUNTER = COUNTER(0,52)
WHITE_OAK_ACORN_MAST = IF (WHITE_OAK_AMP_RATE < 0) THEN 0
ELSE WHITE_OAK_PROPORTION*WHITE_OAK_AMP_RATE
WHITE_OAK_AMP_RATE = IF (THREE_YEAR_COUNTER=129) THEN
(NORMAL(2441,1526)) ELSE IF (THREE_YEAR_COUNTER<>129) AND
(WEEK_COUNTER = 25) THEN (NORMAL(700,480)) ELSE 0
WHITE_OAK_PROPORTION = 1
ACORN_MAST_INDEX = GRAPH(TOTAL_ACORN_MAST)
(0.00, 0.00), (1000, 20.0), (2000, 40.0), (3000, 60.0), (4000, 80.0), (5000, 100),
(6000, 120), (7000, 140), (8000, 160), (9000, 180), (10000, 200)

Tick Burden on Deer
TICK_PER_DEER_PER_DAY = IF (WEEK_COUNTER>=27) AND
(WEEK_COUNTER<=38) THEN 30 ELSE 0

WEEKLY_ADULT_TICK_BURDEN_DEER = DEER_MODEL_FOREST*
TICK_PER_DEER_PER_DAY*(7/7)

Tick Burden on Mice
WEEKLY_LARVAL_TICK_BURDEN =
N*WEEKLY_LARVAL_TICK_BURDEN__MICE
WEEKLY_LARVAL_TICK_BURDEN__MICE =
DAILY_LARVAL_TICK_BURDEN* (7/4) {convert successful biting period per
day to week}
WEEKLY_NYMPH_TICK_BURDEN =
N*WEEKLY_NYMPH_TICK_BURDEN_MICE
WEEKLY_NYMPH_TICK_BURDEN_MICE =
DAILY_NYMPH_TICK_BURDEN * (7/4.5) {convert successful biting period
per day to week}
DAILY_LARVAL_TICK_BURDEN = GRAPH(WEEK_COUNTER)
(0.00, 0.00), (2.00, 3.33), (4.00, 3.33), (6.00, 4.80), (8.00, 4.80), (10.0, 9.90),
(12.0, 9.90), (14.0, 9.90), (16.0, 5.30), (18.0, 5.30), (20.0, 8.70), (22.0, 8.70),
(24.0, 4.40), (26.0, 4.40), (28.0, 1.80), (30.0, 1.80), (32.0, 0.00), (34.0, 0.00),
(36.0, 0.00), (38.0, 0.00), (40.0, 0.00), (42.0, 0.00), (44.0, 0.00), (46.0, 0.00),
(48.0, 0.00), (50.0, 0.00), (52.0, 0.00)
DAILY_NYMPH_TICK_BURDEN = GRAPH(WEEK_COUNTER)
(0.00, 1.35), (2.00, 2.67), (4.00, 2.67), (6.00, 1.10), (8.00, 1.10), (10.0, 0.6),
(12.0, 0.6), (14.0, 0.6), (16.0, 0.3), (18.0, 0.3), (20.0, 0.1), (22.0, 0.1), (24.0, 0.1),
(26.0, 0.1), (28.0, 0.1), (30.0, 0.1), (32.0, 0.00), (34.0, 0.00), (36.0, 0.00), (38.0,
0.00), (40.0, 0.00), (42.0, 0.00), (44.0, 0.00), (46.0, 0.00), (48.0, 0.00), (50.0,
0.00), (52.0, 0.00)

Tick Population Dynamics
ADULT_FED(t) = ADULT_FED(t − dt) + (ADULT_FEED_RATE −
ADULT_REST_RATE) * dt
INIT ADULT_FED = 0
INFLOWS:
ADULT_FEED_RATE =
if (WEEKLY_ADULT_TICK_BURDEN_DEER>=ADULT_UNFED) then
WEEKLY_ADULT_TICK_BURDEN_DEER else (ADULT_UNFED)
OUTFLOWS:
ADULT_REST_RATE = ADULT_FED

ADULT_OVIPOSITION(t) = ADULT_OVIPOSITION(t − dt) +
(ADULT_EMERGENCE_RATE − ADULT_OVIPOSITION_RATE) * dt
INIT ADULT_OVIPOSITION = 300
INFLOWS:
ADULT_EMERGENCE_RATE = CONVEYOR OUTFLOW
OUTFLOWS:
ADULT_OVIPOSITION_RATE = ADULT_OVIPOSITION
ADULT_REST(t) = ADULT_REST(t − dt) + (ADULT_REST_RATE −

ADULT_EMERGENCE_RATE) * dt
INIT ADULT_REST = 0
  TRANSIT TIME = 18
  INFLOW LIMIT = ∞
  CAPACITY = ∞

INFLOWS:
ADULT_REST_RATE = ADULT_FED
OUTFLOWS:
ADULT_EMERGENCE_RATE = CONVEYOR OUTFLOW

ADULT_UNFED(t) = ADULT_UNFED(t − dt) + (N_MOLT −
ADULT_FEED_RATE − ADULT_UNFED_DEATH) * dt
INIT ADULT_UNFED = 0
INFLOWS:
N_MOLT = CONVEYOR OUTFLOW
OUTFLOWS:
ADULT_FEED_RATE =
IF (WEEKLY_ADULT_TICK_BURDEN_DEER>=ADULT_UNFED) THEN
WEEKLY_ADULT_TICK_BURDEN_DEER ELSE (ADULT_UNFED)
ADULT_UNFED_DEATH = (1-ADULT_SURVIVAL_RATE) *
ADULT_UNFED

EGGS(t) = EGGS(t − dt) + (EGG_EMERGENCE_RATE − EGG_HATCH −
EGG_FAILURE) * dt
INIT EGGS = 0
INFLOWS:
EGG_EMERGENCE_RATE = CONVEYOR OUTFLOW
OUTFLOWS:
EGG_HATCH =
(EGGS*EGG_SURVIVAL*EGG_ENVIRONMENTAL_FACTOR_2)
EGG_FAILURE = IF (EGG_ENVIRONMENTAL_FACTOR_2=0) THEN EGGS
ELSE (1 − EGG_SURVIVAL)*EGGS

EGGS_REST(t) = EGGS_REST(t − dt) + (EGG_OVIPOSITION −
EGG_EMERGENCE_RATE − EGG_REST_FAILURE) * dt
INIT EGGS_REST = 0
  TRANSIT TIME = 9
  INFLOW LIMIT = ∞
  CAPACITY = ∞

INFLOWS:
EGG_OVIPOSITION =
ADULT_OVIPOSITION*ADULT_FEMALE_POPULATION_RATIO*
EGG_MAX
OUTFLOWS:
EGG_EMERGENCE_RATE = CONVEYOR OUTFLOW

EGG_REST_FAILURE = LEAKAGE OUTFLOW
LEAKAGE FRACTION = IF (EGG_ENVIRONMENTAL_FACTOR_2=0)
THEN 1 ELSE (1-EGG_FECUNDITY)
NO-LEAK ZONE = 0

LARVAE_ACTIVE(t) = LARVAE_ACTIVE(t − dt) + (EGG_HATCH −
LARVAE_DEATH − LARVAE_FEED) * dt
INIT LARVAE_ACTIVE = 0
INFLOWS:
EGG_HATCH =
(EGGS*EGG_SURVIVAL*EGG_ENVIRONMENTAL_FACTOR_2)
OUTFLOWS:
LARVAE_DEATH = IF (WEEKLY_LARVAL_TICK_BURDEN=0) THEN
LARVAE_ACTIVE ELSE (1-LARVAE_SURV_RATE)*LARVAE_ACTIVE
LARVAE_FEED = IF (LARVAE_ACTIVE >=
WEEKLY_LARVAL_TICK_BURDEN) THEN
WEEKLY_LARVAL_TICK_BURDEN ELSE (LARVAE_ACTIVE)

LARVAE_FED(t) = LARVAE_FED(t −
dt) + (LARVAE_FEED − LARVAE_MOLT − LARVAE_FED_DEATH) * dt
INIT LARVAE_FED = 0
  TRANSIT TIME = 40
  INFLOW LIMIT = ∞
  CAPACITY = ∞

INFLOWS:
LARVAE_FEED = IF (LARVAE_ACTIVE >=
WEEKLY_LARVAL_TICK_BURDEN) THEN
WEEKLY_LARVAL_TICK_BURDEN ELSE (LARVAE_ACTIVE)
OUTFLOWS:
LARVAE_MOLT = CONVEYOR OUTFLOW
LARVAE_FED_DEATH = LEAKAGE OUTFLOW
LEAKAGE FRACTION = (1−LARVAE_SURV_RATE)
NO-LEAK ZONE = 0

NYMPHS_ACTIVE(t) = NYMPHS_ACTIVE(t − dt) + (LARVAE_MOLT −
N_DEATH − N_FEED) * dt
INIT NYMPHS_ACTIVE = 0
INFLOWS:
LARVAE_MOLT = CONVEYOR OUTFLOW
OUTFLOWS:
N_DEATH = IF (WEEKLY_NYMPH_TICK_BURDEN=0) THEN
NYMPHS_ACTIVE ELSE (1−NYMPH_SURVIVAL_RATE) *
NYMPHS_ACTIVE
N_FEED = IF (NYMPHS_ACTIVE >= WEEKLY_NYMPH_TICK_BURDEN)
THEN WEEKLY_NYMPH_TICK_BURDEN ELSE (NYMPHS_ACTIVE)

NYMPH_FED(t) = NYMPH_FED(t − dt) + (N_FEED − N_MOLT -
NYMPH_FED_DEATH) * dt
INIT NYMPH_FED = 0
  TRANSIT TIME = 26
  INFLOW LIMIT = ∞
  CAPACITY = ∞

INFLOWS:
N_FEED = IF (NYMPHS_ACTIVE >= WEEKLY_NYMPH_TICK_BURDEN)
THEN WEEKLY_NYMPH_TICK_BURDEN ELSE (NYMPHS_ACTIVE)
OUTFLOWS:
N_MOLT = CONVEYOR OUTFLOW
NYMPH_FED_DEATH = LEAKAGE OUTFLOW
LEAKAGE FRACTION = (1-NYMPH_SURVIVAL_RATE)
NO-LEAK ZONE = 0

TOTAL_ACTIVE_NYMPHS_YEAR(t) =
TOTAL_ACTIVE_NYMPHS_YEAR(t − dt) +
(TOTAL_ACTIVE_NYMPHS_IN − TOTAL_ACTIVE_NYMPHS_OUT) * dt
INIT TOTAL_ACTIVE_NYMPHS_YEAR = 0
INFLOWS:
TOTAL_ACTIVE_NYMPHS_IN = LARVAE_MOLT
OUTFLOWS:
TOTAL_ACTIVE_NYMPHS_OUT = IF LARVAE_MOLT = 0 THEN
TOTAL_ACTIVE_NYMPHS_YEAR ELSE 0

ADULT_FEMALE_POPULATION_RATIO = 0.5
ADULT_SURVIVAL_RATE = .92
EGG_ENVIRONMENTAL_FACTOR = (SIN(2*PI/52*TIME))*1.632
EGG_ENVIRONMENTAL_FACTOR_2 =
IF (EGG_ENVIRONMENTAL_FACTOR<0) THEN 0 ELSE
EGG_ENVIRONMENTAL_FACTOR
EGG_FECUNDITY = 1
EGG_MAX = 3000 {Max number of eggs per oviposition}
EGG_SURVIVAL = .0336
LARVAE_SURV_RATE = .37
NYMPH_SURVIVAL_RATE = .92

Tick Population Dynamics 2
AADULT_FED_2(t) = AADULT_FED_2(t − dt) + (ADULT_FEED_2 −
ADULT_REST_RATE_2) * dt
INIT AADULT_FED_2 = 0
INFLOWS:
ADULT_FEED_2 =
IF (WEEKLY_ADULT_TICK_BURDEN_DEER>=ADULT_UNFED_2)
THEN WEEKLY_ADULT_TICK_BURDEN_DEER ELSE
(ADULT_UNFED_2)

OUTFLOWS:
ADULT_REST_RATE_2 = AADULT_FED_2

ADULT_OVIPOSITION_2(t) = ADULT_OVIPOSITION_2(t − dt) +
(ADULT_EMERGENCE_RATE_2 − ADULT_OVIPOSITION_DEATH_2) * dt
INIT ADULT_OVIPOSITION_2 = 0
INFLOWS:
ADULT_EMERGENCE_RATE_2 = CONVEYOR OUTFLOW
OUTFLOWS:
ADULT_OVIPOSITION_DEATH_2 = ADULT_OVIPOSITION_2

ADULT_REST_2(t) = ADULT_REST_2(t − dt) + (ADULT_REST_RATE_2 −
ADULT_EMERGENCE_RATE_2) * dt
INIT ADULT_REST_2 = 0
   TRANSIT TIME = 18
   INFLOW LIMIT = ∞
   CAPACITY = ∞

INFLOWS:
ADULT_REST_RATE_2 = AADULT_FED_2
OUTFLOWS:
ADULT_EMERGENCE_RATE_2 = CONVEYOR OUTFLOW

ADULT_UNFED_2(t) = ADULT_UNFED_2(t − dt) + (NYMPH_MOLT_2 −
ADULT_FEED_2 − ADULT_UNFED_DEATH_2) * dt
INIT ADULT_UNFED_2 = 0
INFLOWS:
NYMPH_MOLT_2 = CONVEYOR OUTFLOW
OUTFLOWS:
ADULT_FEED_2 =
IF (WEEKLY_ADULT_TICK_BURDEN_DEER>=ADULT_UNFED_2)
THEN WEEKLY_ADULT_TICK_BURDEN_DEER ELSE
(ADULT_UNFED_2)
ADULT_UNFED_DEATH_2 = (1-ADULT_SURVIVAL_RATE) *
ADULT_UNFED_2

EGGS_2(t) = EGGS_2(t − dt) + (EGG_EMERGENCE_RATE_2 −
EGG_HATCH_2 − EGG_FAILURE_2) * dt
INIT EGGS_2 = 0
INFLOWS:
EGG_EMERGENCE_RATE_2 = CONVEYOR OUTFLOW
OUTFLOWS:
EGG_HATCH_2 =
(EGGS_2*EGG_SURVIVAL*EGG_ENVIRONMENTAL_FACTOR_2)
EGG_FAILURE_2 = IF (EGG_ENVIRONMENTAL_FACTOR_2=0) THEN
EGGS_2 ELSE (1-EGG_SURVIVAL)*EGGS_2

EGG_REST_2(t) = EGG_REST_2(t − dt) + (EGG_OVIPOSITION_2 −

EGG_EMERGENCE_RATE_2 − EGG_REST_FAILURE_2) * dt
INIT EGG_REST_2 = 0
   TRANSIT TIME = 9
   INFLOW LIMIT = ∞
   CAPACITY = ∞

INFLOWS:
EGG_OVIPOSITION_2 = ADULT_OVIPOSITION_2*
ADULT_FEMALE_POPULATION_RATIO*EGG_MAX
OUTFLOWS:
EGG_EMERGENCE_RATE_2 = CONVEYOR OUTFLOW
EGG_REST_FAILURE_2 = LEAKAGE OUTFLOW
LEAKAGE FRACTION = IF (EGG_ENVIRONMENTAL_FACTOR_2=0)
THEN 1 ELSE (1−EGG_FECUNDITY)
NO-LEAK ZONE = 0

LARVAE_ACTIVE_2(t) = LARVAE_ACTIVE_2(t − dt) + (EGG_HATCH_2 −
L_DEATH_2 − LARVAE_FEED_2) * dt
INIT LARVAE_ACTIVE_2 = 0
INFLOWS:
EGG_HATCH_2 =
(EGGS_2*EGG_SURVIVAL*EGG_ENVIRONMENTAL_FACTOR_2)
OUTFLOWS:
L_DEATH_2 = IF (WEEKLY_LARVAL_TICK_BURDEN=0) THEN
LARVAE_ACTIVE_2 ELSE
(1-LARVAE_SURV_RATE)*LARVAE_ACTIVE_2
LARVAE_FEED_2 = IF (LARVAE_ACTIVE_2 >=
WEEKLY_LARVAL_TICK_BURDEN) THEN
WEEKLY_LARVAL_TICK_BURDEN ELSE (LARVAE_ACTIVE_2)

LARVAE_FED_2(t) = LARVAE_FED_2(t − dt) + (LARVAE_FEED_2 −
LARVAE_MOLT_2 − LARVAE_FED_DEATH_2) * dt
INIT LARVAE_FED_2 = 0
   TRANSIT TIME = 40
   INFLOW LIMIT = ∞
   CAPACITY = ∞

INFLOWS:
LARVAE_FEED_2 = IF (LARVAE_ACTIVE_2 >=
WEEKLY_LARVAL_TICK_BURDEN) THEN
WEEKLY_LARVAL_TICK_BURDEN ELSE (LARVAE_ACTIVE_2)
OUTFLOWS:
LARVAE_MOLT_2 = CONVEYOR OUTFLOW
LARVAE_FED_DEATH_2 = LEAKAGE OUTFLOW
LEAKAGE FRACTION = (1−LARVAE_SURV_RATE)
NO-LEAK ZONE = 0

NYMPH_ACTIVE_2(t) = NYMPH_ACTIVE_2(t − dt) + (LARVAE_MOLT_2 −
N_DEATH_2 − NYMPH_FEED_2) * dt
INIT NYMPH_ACTIVE_2 = 350
INFLOWS:
LARVAE_MOLT_2 = CONVEYOR OUTFLOW
OUTFLOWS:
N_DEATH_2 = IF (WEEKLY_NYMPH_TICK_BURDEN=0) THEN
NYMPH_ACTIVE_2 ELSE (1-NYMPH_SURVIVAL_RATE) *
NYMPH_ACTIVE_2
NYMPH_FEED_2 = IF (NYMPH_ACTIVE_2 >=
WEEKLY_NYMPH_TICK_BURDEN) THEN
WEEKLY_NYMPH_TICK_BURDEN ELSE (NYMPH_ACTIVE_2)

NYMPH_FED_2(t) = NYMPH_FED_2(t − dt) + (NYMPH_FEED_2 −
NYMPH_MOLT_2 − NYMPH_FED_DEATH_2) * dt
INIT NYMPH_FED_2 = 0
   TRANSIT TIME = 26
   INFLOW LIMIT = ∞
   CAPACITY = ∞

INFLOWS:
NYMPH_FEED_2 = IF (NYMPH_ACTIVE_2 >=
WEEKLY_NYMPH_TICK_BURDEN) THEN
WEEKLY_NYMPH_TICK_BURDEN ELSE (NYMPH_ACTIVE_2)
OUTFLOWS:
NYMPH_MOLT_2 = CONVEYOR OUTFLOW
NYMPH_FED_DEATH_2 = LEAKAGE OUTFLOW
LEAKAGE FRACTION = (1-NYMPH_SURVIVAL_RATE)
NO-LEAK ZONE = 0

TOTAL_ACTIVE_NYMPHS_YEAR_2(t) =
TOTAL_ACTIVE_NYMPHS_YEAR_2(t − dt) +
(TOTAL_ACTIVE_NYMPHS_IN_2 −
TOTAL_ACTIVE_NYMPHS_OUT_2) * dt
INIT TOTAL_ACTIVE_NYMPHS_YEAR_2 = 0
INFLOWS:
TOTAL_ACTIVE_NYMPHS_IN_2 = LARVAE_MOLT_2
OUTFLOWS:
TOTAL_ACTIVE_NYMPHS_OUT_2 = IF LARVAE_MOLT_2 = 0 THEN
TOTAL_ACTIVE_NYMPHS_YEAR_2 ELSE 0

Chapter 8
Chicken Pox and Shingles

Chicken pox, a microparasitic infection caused by the *varicella zoster* virus, is an example of a highly infectious childhood disease. This virus can be spread either through direct contact with infected individuals or through the air. After exposure to the virus, the incubation period before an individual becomes contagious is approximately seven days. Individuals are contagious for about seven days during which symptoms including fever and blisters appear, and then remain sick for an additional fourteen days. Once an individual recovers, he or she develops a natural immunity and is unlikely to get the disease again.

However, the virus remains in the body and later in life, it manifests itself in the form of shingles in about 15 percent of the population that contracted chicken pox. Shingles has symptoms that are similar to chicken pox but strikes mostly individuals over the age of 50 that are fatigued or under stress. It takes approximately 10 days to recover from shingles and during this time susceptible individuals can contract chicken pox from those suffering from shingles.

New vaccines are becoming available to immunize people against chicken pox. The target population for immunization typically is children, because they comprise the highest infectious class. Immunizing children would help to reduce occurrence of chicken pox. However, no vaccination program can reliably cover 100 percent of a population. Consequently, as childhood vaccination takes place, fewer cases of chicken pox among children occur but the age of first infection increases. Since older individuals have more difficulty dealing with chicken pox than young individuals, vaccination programs may shift the burden from the young to the old.

The following model addresses several interrelated questions: How are different population age cohorts affected by chicken pox in the absence of immunization and shingles? What are the effects of immunization on the average age of first contraction of chicken pox? What are the effects of shingles on the average age of first contraction?

To answer these questions, we first explore the effects of chicken pox on different age cohorts in a given population. We then investigate what happens to the same population when the effects of shingles are factored into the model. Finally, we turn to the effects of childhood immunization on the average age at which a person contracts the disease.

B. Hannon and M. Ruth, *Dynamic Modeling of Diseases and Pests*,
Modeling Dynamic Systems,
© Springer Science + Business Media LLC 2009

8.1 Model Assumptions and Structure

For illustrative purposes, we initialize our model with population data from the 1990 U.S. Census and group them into six age cohorts. The first five cohorts span 10 years (3,650 days) each, and the last comprises the population of 50 years and older. By assumption, individuals in the last cohort remain a maximum of 30 years (10,950 days) in that cohort. The total initial population is 245,704, measured in thousands of individuals. For simplicity, we assume that only 50 percent of individuals in the 10- to 19-year age cohort may reproduce, and that the total reproductive population includes them as well as all individuals in the 20- to 39-year cohort. Further, we assume a uniform and constant death rate of .000027 per day.

Following research on infections with chicken pox[1], the average number of infections with chicken pox by an infected individual is assumed to be

$$RsubZero = 10 \tag{8.1}$$

and individuals are removed from the infective stock at a rate of

$$V = 1/7. \tag{8.2}$$

The general form for the transmission coefficient is[2]

$$Beta = RsubZero * V/Population. \tag{8.3}$$

With an initial population of 245,704 and V = 1/7, Beta is .0000058.
A general representation of the transmission rate is

$$Transmission\ Rate = Beta * Total\ Infective * Susceptible \tag{8.4}$$

Since transmission rates for chicken pox vary among age cohorts, we weight the transmission coefficient for different age cohorts:

$$Transmission\ Rate = (Beta\ Weight * Beta) * Total\ Infective * Susceptible \tag{8.5}$$

Individuals are most likely to contract chicken pox when they are young. This is because children are typically kept in close contact with each other in places such as school and daycare. We choose a Beta Weight of 1 for the first age cohort, thus making the virus have its full effect in transmission. As people get older, however, they are not as likely to come into contact with the virus. Therefore, lower weights are used. The weight for the 10- to 19-year age cohort is set at .8. These individuals continue to be in close contact with each other, but a large percentage of the

---

[1] Hethcote, H.W. "Qualitative analyses of communicable disease models." Mathematical Biosciences from *Mathematical Models in Biology*.
    May, Robert M. 1983."Parasitic Infections as Regulator of Animal Populations." American Scientist. 71: 36–45 in *Mathematical Models in Biology*.

[2] Edelstein-Keshet, Leah. 1988. *Mathematical Models in Biology*. Random House: New York.

individuals they come into contact with are already immune to chicken pox. For the four older age cohorts, the weights are decreased, taking on values of 0.5 for the 10- to 19-year cohort, 0.25 for the 20- to 29-year cohort, 0.1 for the 30- to 39-year cohort and 0.05 for the 50-year cohort.

We assume that the virus is present at the beginning of the model run. Specifically, we postulate that 20 percent of the population have contracted chicken pox by the age of 4; 65 percent by the age of 9; 88 percent by the age of 19; 95 percent by the age of 29; 97 percent by the age of 39; 98 percent by the age of 49; and 99 percent by age 50 or older[3]. The initial susceptible population in each age cohort, measured in thousands of individuals, is shown in Table 8.1. Initially, no individual in either age cohort is assumed to be within the incubation period, and the incubation time for all age cohorts is 7 days.

Initial infective stocks (measured in thousands of individuals) are given in Table 8.1. Since chicken pox is mainly a childhood disease, the assumption is made that the largest infective stock would be Infective 1, for the age cohort of 0 to 9 years. The infective stock was then decreased for each subsequent age cohort.

The total number of infective individuals includes those who are infected with shingles. By assumption, only individuals in the last age cohort may be afflicted with shingles, and at the beginning of the model no one suffers from shingles. To model the contraction with shingles, we invoke a "shingle rate" of 0, which allows us to study its effects when increased. For subsequent runs, we assume 0.15 cases of shingles per thousand individuals, and a recovery time of 10 days. In contrast, the number of days it takes individual to recover from chicken pox is 14.

The model is shown in Figure 8.1. It contains a flow to reflect immunization, calculated as the product of the number of individuals who are susceptible to chicken pox and an immunization rate. For the model runs discussed below, we varied immunization rates, starting at zero and increasing it for subsequent runs to reflect more prevalent childhood immunization. The total stock of infective individuals is shown as the sum of all individuals with chicken pox and with shingles. The total force of infection is simply defined as the (unweighted) transmission coefficient multiplied by the total number of infective individuals. From this, we can calculate the average age of first infection as the inverse of the force of infection, divided by 365 days.

Table 8.1 Initial Values of Stocks (Thousands of Individuals)

Cohort	Susceptibles	Infective	Immune
0–9	20,710	40	15,252
10–19	93,961	8	29,035
20–29	1927	1	36,604
30–39	1254	1	40,541
40–49	654	0	32,044
50 Plus	637	0	63,033

[3] Finger, Reginald, Jeffrey Hughes, Barry J. Meade, Andrew R. Pelletier, and Clarkson T. Palmer. "Age-Specific Incidence of Chickenpox." Public Health Reports. Nov/Dec 1994: 750–755.

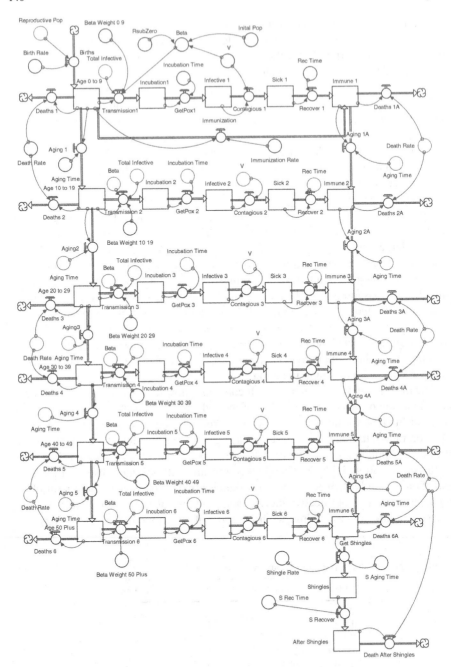

Fig. 8.1

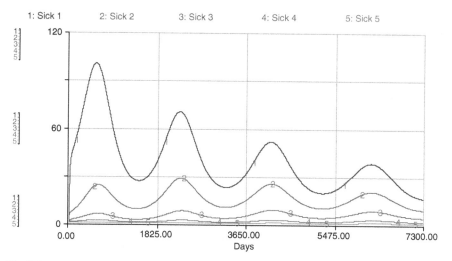

Fig. 8.2

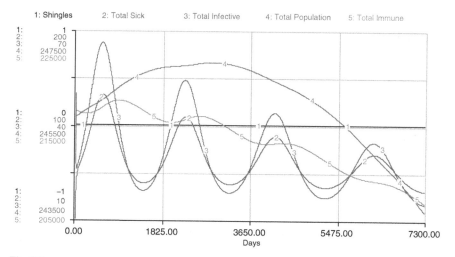

Fig. 8.3

Model results without immunization and without the effects of shingles (i.e. the "shingle rate" set to zero instead of 0.15), are shown in Figure 8.2 for the first five cohorts. The sixth cohort—individuals of 50 years and older—are virtually unaffected by chicken pox over the simulated 20-yeartimeframe. Overall, the number of individuals sick from chicken pox declines, in part because the population in the long run is declining (Figure 8.3). As the population ages and no vaccination takes place, older parts of the population are relatively more affected by chicken pox (Figure 8.3), and the average age of first infection increases (Figure 8.4).

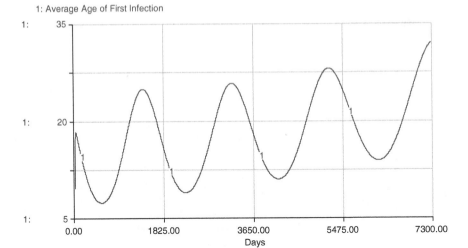

Fig. 8.4

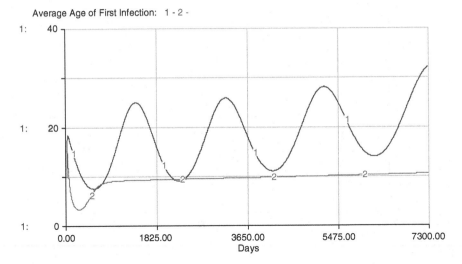

Fig. 8.5

Figure 8.5 compares the previous results where there are no shingles in the population to the case where the elderly are affected by shingles. The presence of shingles
leads to less pronounced impacts of the disease on the average age of infection and
clearly alleviates the cyclical nature of infection in the population as a whole. These
results are the consequence of having a larger infective population that can pass the
virus to young and old, and of having different timeframes during which the virus
can be passed to the susceptible population.

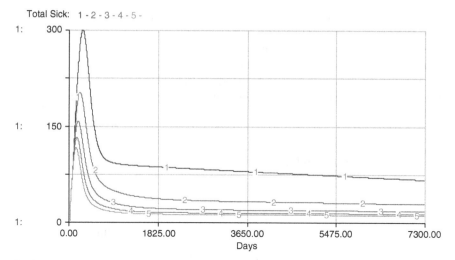

Fig. 8.6

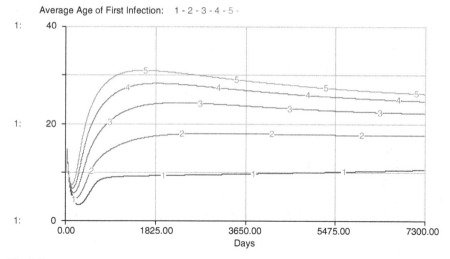

Fig. 8.7

The results shown in Figures 8.6 and 8.7 compare the effects of four alternative immunization rates—0, 0.0005, 0.001, 0.0015, and 0.002. In all cases, we consider also the effects of shingles on the spread of chicken pox. Higher immunization rates reduce the number of people getting sick, but higher immunization rates also have higher ages of first infection associated with them.

Although fewer people overall will get chicken pox, the health risks are greater for those who contract it because they will be older[4]. Chicken pox can be especially

---

[4] Wolinsky, Howard. "Its Effect on Shingles Is Studied." *Chicago Sun-Times*. May 7, 1995: 62+.

dangerous for pregnant women and their unborn children. Pregnant women are likely to be in the same age bracket as the average age of first infection under the three more aggressive immunization policies shown as lines 2, 3, and 4 in Figure 8.7. Even with a less aggressive immunization policy, the average age at which individuals contract the disease is still high enough to cause a health risk to individuals. Therefore, the establishment of a vaccination policy for chicken pox must be given careful consideration.

8.2 Questions and Tasks

1. Explore the effects of different Beta Weights on model results. Choose larger weights for younger age cohorts.
2. Explore the effects of different vaccination rates on model results.

   a. Choose the same vaccination rate for the entire population.
   b. Choose different vaccination rate for different age cohorts.

3. Assume an increasing total population and determine the effects of the dynamics of chicken pox, especially the average age of first infections.

CHICKEN POX AND SHINGLES

After_Shingles(t) = After_Shingles(t − dt) + (S_Recover − Death_After_Shingles) * dt
INIT After_Shingles = 0

INFLOWS:
S_Recover = Shingles/S_Rec_Time {people per day}
OUTFLOWS:
Death_After_Shingles = After_Shingles*Death_Rate
Age_0_to_9(t) = Age_0_to_9(t − dt) + (Births - Aging_1 − Transmission1 − Immunization − Deaths_1) * dt
INIT Age_0_to_9 = 20710 {people}

INFLOWS:
Births = Birth_Rate*Reproductive_Pop {births per day}
OUTFLOWS:
Aging_1 = Age_0_to_9/Aging_Time {people per day}
Transmission1 = (Beta_Weight_0_9*Beta)*Age_0_to_9*Total_Infective {people exposed per day}
Immunization = Age_0_to_9*Immunization_Rate {people per population per day}
Deaths_1 = Age_0_to_9*Death_Rate {deaths per day}
Age_10_to_19(t) = Age_10_to_19(t − dt) + (Aging_1 − Transmission_2 − Aging2 − Deaths_2) * dt
INIT Age_10_to_19 = 3961 {people}

INFLOWS:

Aging_1 = Age_0_to_9/Aging_Time {people per day}

OUTFLOWS:

Transmission_2 = (Beta_Weight_10_19*Beta)*Age_10_to_19*Total_Infective {people exposed per day}

Aging2 = Age_10_to_19/Aging_Time {people per day}

Deaths_2 = Age_10_to_19*Death_Rate {deaths per day}

Age_20_to_29(t) = Age_20_to_29(t − dt) + (Aging2 − Transmission_3 − Aging3 − Deaths_3) * dt

INIT Age_20_to_29 = 1927 {people}

INFLOWS:

Aging2 = Age_10_to_19/Aging_Time {people per day}

OUTFLOWS:

Transmission_3 = (Beta_Weight_20_29*Beta)*Age_20_to_29*Total_Infective {people exposed per day}

Aging3 = Age_20_to_29/Aging_Time {people per day}

Deaths_3 = Age_20_to_29*Death_Rate {deaths per day}

Age_30_to_39(t) = Age_30_to_39(t − dt) + (Aging3 − Transmission_4 − Aging_4 − Deaths_4) * dt

INIT Age_30_to_39 = 1254 {people}

INFLOWS:

Aging3 = Age_20_to_29/Aging_Time {people per day}

OUTFLOWS:

Transmission_4 = (Beta_Weight_30_39*Beta)*Age_30_to_39*Total_Infective {people exposed per day}

Aging_4 = Age_30_to_39/Aging_Time {people per day}

Deaths_4 = Age_30_to_39*Death_Rate {deaths per day}

Age_40_to_49(t) = Age_40_to_49(t − dt) + (Aging_4 − Transmission_5 − Aging_5 − Deaths_5) * dt

INIT Age_40_to_49 = 654 {people}

INFLOWS:

Aging_4 = Age_30_to_39/Aging_Time {people per day}

OUTFLOWS:

Transmission_5 = (Beta_Weight_40_49*Beta)*Age_40_to_49*Total_Infective {people exposed per day}

Aging_5 = Age_40_to_49/Aging_Time {people per day}

Deaths_5 = Age_40_to_49*Death_Rate {deaths per day}

Age_50_Plus(t) = Age_50_Plus(t − dt) + (Aging_5 - Transmission_6 − Deaths_6) * dt

INIT Age_50_Plus = 637 {people}

INFLOWS:

Aging_5 = Age_40_to_49/Aging_Time {people per day}

OUTFLOWS:

Transmission_6 = (Beta_Weight_50_Plus*Beta)*Age_50_Plus*Total_Infective {people exposed per day}

Deaths_6 = Age_50_Plus*Death_Rate {deaths per day}

Immune_1(t) = Immune_1(t − dt) + (Recover_1 + Immunization − Aging_1A − Deaths_1A) * dt

INIT Immune_1 = 15252 {people}

INFLOWS:

Recover_1 = Sick_1/Rec_Time {people recovered per day}

Immunization = Age_0_to_9*Immunization_Rate {people per population per day}

OUTFLOWS:

Aging_1A = Immune_1/Aging_Time {people per day}

Deaths_1A = Immune_1*Death_Rate {deaths per day}

Immune_2(t) = Immune_2(t − dt) + (Recover_2 + Aging_1A − Aging_2A − Deaths_2A) * dt

INIT Immune_2 = 29035 {people}

INFLOWS:

Recover_2 = Sick_2/Rec_Time {people recovered per day}

Aging_1A = Immune_1/Aging_Time {people per day}

OUTFLOWS:

Aging_2A = Immune_2/Aging_Time

Deaths_2A = Immune_2*Death_Rate {deaths per day}

Immune_3(t) = Immune_3(t − dt) + (Recover_3 + Aging_2A − Aging_3_A − Deaths_3A) * dt

INIT Immune_3 = 36604 {people}

INFLOWS:

Recover_3 = Sick_3/Rec_Time {people recovered per day}

Aging_2A = Immune_2/Aging_Time

OUTFLOWS:

Aging_3_A = Immune_3/Aging_Time {people per day}

Deaths_3A = Immune_3*Death_Rate {deaths per Day}

Immune_4(t) = Immune_4(t − dt) + (Recover_4 + Aging_3_A − Aging_4A − Deaths_4A) * dt

INIT Immune_4 = 40541 {people}

INFLOWS:

Recover_4 = Sick_4/Rec_Time {people recovered per day}

Aging_3_A = Immune_3/Aging_Time {people per day}

OUTFLOWS:

Aging_4A = Immune_4/Aging_Time {people per day}

Deaths_4A = Immune_4*Death_Rate {deaths per day}

Immune_5(t) = Immune_5(t − dt) + (Recover_5 + Aging_4A − Aging_5A − Deaths_5A) * dt

INIT Immune_5 = 32044 {people}

INFLOWS:
Recover_5 = Sick_5/Rec_Time {people recovered per day}
Aging_4A = Immune_4/Aging_Time {people per day}
OUTFLOWS:
Aging_5A = Immune_5/Aging_Time {people per day}
Deaths_5A = Immune_5*Death_Rate {deaths per day}
Immune_6(t) = Immune_6(t − dt) + (Recover_6 + Aging_5A − Get_Shingles − Deaths_6A) * dt

INIT Immune_6 = 63033 {people}

INFLOWS:
Recover_6 = Sick_6/Rec_Time {people recovered per day}
Aging_5A = Immune_5/Aging_Time {people per day}
OUTFLOWS:
Get_Shingles = Immune_6*Shingle_Rate/S_Aging_Time {people getting shingles per day}
Deaths_6A = Immune_6*Death_Rate {deaths per day}
Incubation1(t) = Incubation1(t − dt) + (Transmission1 − GetPox1) * dt
INIT Incubation1 = 0 {people}

INFLOWS:
Transmission1 = (Beta_Weight_0_9*Beta)*Age_0_to_9*Total_Infective {people exposed per day}
OUTFLOWS:
GetPox1 = Incubation1/Incubation_Time {people breaking out in chicken pox per day}
Incubation_2(t) = Incubation_2(t − dt) + (Transmission_2 − GetPox_2) * dt
INIT Incubation_2 = 0 {people}

INFLOWS:
Transmission_2 = (Beta_Weight_10_19*Beta)*Age_10_to_19*Total_Infective {people exposed per day}
OUTFLOWS:
GetPox_2 = Incubation_2/Incubation_Time {people breaking out in chicken pox per day}
Incubation_3(t) = Incubation_3(t − dt) + (Transmission_3 − GetPox_3) * dt
INIT Incubation_3 = 0 {people}

INFLOWS:
Transmission_3 = (Beta_Weight_20_29*Beta)*Age_20_to_29*Total_Infective {people exposed per day}

OUTFLOWS:

GetPox_3 = Incubation_3/Incubation_Time {people breaking out in chicken pox per day}

Incubation_4(t) = Incubation_4(t − dt) + (Transmission_4 − GetPox_4) * dt

INIT Incubation_4 = 0 {people}

INFLOWS:

Transmission_4 = (Beta_Weight_30_39*Beta)*Age_30_to_39*Total_Infective {people exposed per day}

OUTFLOWS:

GetPox_4 = Incubation_4/Incubation_Time {people breaking out in chicken pox per day}

Incubation_5(t) = Incubation_5(t − dt) + (Transmission_5 − GetPox_5) * dt

INIT Incubation_5 = 0 {people}

INFLOWS:

Transmission_5 = (Beta_Weight_40_49*Beta)*Age_40_to_49*Total_Infective {people exposed per day}

OUTFLOWS:

GetPox_5 = Incubation_5/Incubation_Time {people breaking out in chicken pox per day}

Incubation_6(t) = Incubation_6(t − dt) + (Transmission_6 − GetPox_6) * dt

INIT Incubation_6 = 0 {people}

INFLOWS:

Transmission_6 = (Beta_Weight_50_Plus*Beta)*Age_50_Plus*Total_Infective {people exposed per day}

OUTFLOWS:

GetPox_6 = Incubation_6/Incubation_Time {people breaking out in chicken pox per day}

Infective_1(t) = Infective_1(t − dt) + (GetPox1 − Contagious_1) * dt

INIT Infective_1 = 40 {people}

INFLOWS:

GetPox1 = Incubation1/Incubation_Time {people breaking out in chicken pox per day}

OUTFLOWS:

Contagious_1 = Infective_1*V {people contagious per day}

Infective_2(t) = Infective_2(t − dt) + (GetPox_2 − Contagious_2) * dt

INIT Infective_2 = 8 {people}

INFLOWS:

GetPox_2 = Incubation_2/Incubation_Time {people breaking out in chicken pox per day}

OUTFLOWS:

Contagious_2 = Infective_2*V {people contagious per day}

Infective_3(t) = Infective_3(t − dt) + (GetPox_3 − Contagious_3) * dt

INIT Infective_3 = 1 {people}

INFLOWS:
GetPox_3 = Incubation_3/Incubation_Time {people breaking out in chicken pox per day}
OUTFLOWS:
Contagious_3 = Infective_3*V {people contagious per day}
Infective_4(t) = Infective_4(t − dt) + (GetPox_4 − Contagious_4) * dt
INIT Infective_4 = 1 {people}

INFLOWS:
GetPox_4 = Incubation_4/Incubation_Time {people breaking out in chicken pox per day}
OUTFLOWS:
Contagious_4 = Infective_4*V {people contagious per day}
Infective_5(t) = Infective_5(t − dt) + (GetPox_5 − Contagious_5) * dt
INIT Infective_5 = 0 {people}

INFLOWS:
GetPox_5 = Incubation_5/Incubation_Time {people breaking out in chicken pox per day}
OUTFLOWS:
Contagious_5 = Infective_5*V {people contagious per day}
Infective_6(t) = Infective_6(t − dt) + (GetPox_6 − Contagious_6) * dt
INIT Infective_6 = 0 {people}

INFLOWS:
GetPox_6 = Incubation_6/Incubation_Time {people breaking out in chicken pox per day}
OUTFLOWS:
Contagious_6 = Infective_6*V {people contagious per day}
Shingles(t) = Shingles(t − dt) + (Get_Shingles − S_Recover) * dt
INIT Shingles = 0 {people}

INFLOWS:
Get_Shingles = Immune_6*Shingle_Rate/S_Aging_Time {people getting shingles per day}
OUTFLOWS:
S_Recover = Shingles/S_Rec_Time {people per day}
Sick_1(t) = Sick_1(t − dt) + (Contagious_1 − Recover_1) * dt
INIT Sick_1 = 0 {people}

INFLOWS:
Contagious_1 = Infective_1*V {people contagious per day}
OUTFLOWS:
Recover_1 = Sick_1/Rec_Time {people recovered per day}
Sick_2(t) = Sick_2(t − dt) + (Contagious_2 − Recover_2) * dt
INIT Sick_2 = 0 {people}

INFLOWS:
Contagious_2 = Infective_2*V {people contagious per day}
OUTFLOWS:
Recover_2 = Sick_2/Rec_Time {people recovered per day}
Sick_3(t) = Sick_3(t − dt) + (Contagious_3 − Recover_3) * dt
INIT Sick_3 = 0 {people}

INFLOWS:
Contagious_3 = Infective_3*V {people contagious per day}
OUTFLOWS:
Recover_3 = Sick_3/Rec_Time {people recovered per day}
Sick_4(t) = Sick_4(t − dt) + (Contagious_4 − Recover_4) * dt
INIT Sick_4 = 0 {people}

INFLOWS:
Contagious_4 = Infective_4*V {people contagious per day}
OUTFLOWS:
Recover_4 = Sick_4/Rec_Time {people recovered per day}
Sick_5(t) = Sick_5(t − dt) + (Contagious_5 − Recover_5) * dt
INIT Sick_5 = 0 {people}

INFLOWS:
Contagious_5 = Infective_5*V {people contagious per day}
OUTFLOWS:
Recover_5 = Sick_5/Rec_Time {people recovered per day}
Sick_6(t) = Sick_6(t − dt) + (Contagious_6 − Recover_6) * dt
INIT Sick_6 = 0 {people}

INFLOWS:
Contagious_6 = Infective_6*V {people contagious per day}
OUTFLOWS:
Recover_6 = Sick_6/Rec_Time {people recovered per day}
Aging_Time = 3650 {days}
Average_Age_of_First_Infection = (1/Per_Capita_Force_of_Infection)/365 {years}
Beta = RsubZero*V/Inital_Pop {Transmission Coefficient}
Beta_Weight_0_9 = 1
Beta_Weight_10_19 = .8
Beta_Weight_20_29 = .5
Beta_Weight_30_39 = .25
Beta_Weight_40_49 = .1
Beta_Weight_50_Plus = .05
Birth_Rate = .000078 {births per reproductive population per day}
Death_Rate = .000027 {deaths per population per day}
Immunization_Rate = .000 {children immunized per population per day}
Incubation_Time = 7 {days}
Inital_Pop = 245704 {people}

Per_Capita_Force_of_Infection = Beta*Total_Infective {1/days}

Rec_Time = 14 {days}

Reproductive_Pop = .5*(Age_10_to_19+Incubation_2+Infective_2+Sick_2+
Immune_2)+Age_20_to_29+
Incubation_3+Infective_3+Sick_3+Immune_3+Age_30_to_39+Incubation_4+
Infective_4+Sick_4+Immune_4+Sick_4 {people}

RsubZero = 10 {average number of infections caused by one infected individual}

Shingle_Rate = .15 {cases of shingles per population per day}

S_Aging_Time = 10950 {days}

S_Rec_Time = 10 {days}

Total_Immune = Immune_1 + Immune_2 + Immune_3 + Immune_4 +
Immune_5 + Immune_6

Total_Infective = Infective_1+Infective_2+Infective_3+Infective_4+Infective_5+
Infective_6+Shingles {people}

Total_Population = Age_0_to_9 + Age_10_to_19 + Age_20_to_29 +
Age_30_to_39 + Age_40_to_49 + Age_50_Plus + Immune_1 + Immune_2 +
Immune_3 + Immune_4 + Immune_5 + Immune_6 + Incubation1 +
Incubation_2 + Incubation_3 + Incubation_4 + Incubation_5 + Incubation_6 +
Infective_1 + Infective_2 + Infective_3 + Infective_4 + Infective_5 + Infective_6
+ Shingles + Sick_3 + Sick_4 + Sick_5 + Sick_6

Total_Sick = Sick_1 + Sick_2 + Sick_3 + Sick_4 + Sick_5 + Sick_6

V = 1/7 {removal rate from the infective stock (1/time)}

Chapter 9
Toxoplasmosis[1]

9.1 Introduction

Toxoplasmosis is a parasitic infection caused by the protozoan *Toxoplasma gondii*. Humans can become infected by ingestion of raw or undercooked meat that contains tissue oocysts (the reproductive cell), or by direct contamination from the environment contaminated by infected cat feces. Although toxoplasmosis is most often asymptomatic in humans, it can cause serious illness in immune-compromised individuals and in fetuses. At this time, there is no cure for the disease; prevention is the only method of control.

This model examines the principal factors that influence the spread of the disease from animals to humans on a case study of 43 Illinois swine farms. Most of the data for this model were obtained from[2,3], and[4]. Some of the references are given in the model variable icons.

The main questions we wish to answer with this model are: How is the prevalence of *Toxoplasma gondii* in people on swine farms in Illinois affected by exposure to cats, ingestion of infected food, and the handling of dirt? Which one of these factors affects the prevalence most? How would the prevalence be affected if we vaccinated the cats, and which rate would be optimal?

---

[1] This chapter is condensed from the work of Nohra Mateus-Pinilla, Illinois Natural History Survey, Champaign, IL.

[2] Frenkel, J.K. et al., 1981. "Endemicity of Toxoplasmosis in Costa Rica," *Am. J. Epidemiology,* v113:254–269.

[3] Smith, J.L 1993. "Documented Outbreaks of Toxoplasmosis: Transmission of *Toxoplasma gondii* to Humans," *J. Food Protection,* v56:630–639.

[4] Weigel, et al., 1997. "Risk Factors for Infection with *Toxoplasma gondii* for Residents and Workers on Swine Farms in Illinois," Vet. Biosciences Dept., University of Illinois, Urbana IL, unpublished.

B. Hannon and M. Ruth, *Dynamic Modeling of Diseases and Pests,*
Modeling Dynamic Systems,
© Springer Science + Business Media LLC 2009

9.2 Model Construction

The life cycle of the *T. gondii* parasite has three stages: cyst, oocyst, or tachyzoite. A cat may eat cysts in infected rodents or birds, or other raw meat; then the organisms will begin to multiply in the wall of the small intestine, producing the second stage, oocysts. These are excreted in the feces for 2 to 3 weeks. Then they may become spores and become infectious to other animals, including humans. Most exposed cats shed oocysts during acute Toxoplasma infection, but not after. Oocysts are very hardy and can survive in moist, shaded soil or sand for months. They can be passed on directly to animals and humans, as well as indirectly to humans who consume meat that is undercooked.

To capture the toxoplasmosis dynamics, we need information on the growth and infection of the human population (by gender), the cat population, and the rat population. Total human population on these farms was 174, (77% male). We begin our model with knowledge that the mean prevalence of the disease in humans is a surprising 31%. We assume the human birth rate is a normal distribution, centered on a mean of 16 births per 1,000 residents. The natural human death rate is also a normal distribution, centered on a mean of 0.85%. Infected humans have a higher death rate of 2%. The infection rate for female humans is described by:

$$\text{Human IR} = (\text{Percent\_Raw\_Food} * \text{Pig\_Prev} * 1) + (\text{Inf\_Cat\_Density} * 0.003)$$
$$+ (\text{Dirt\_Handling} * 0.005), \tag{9.1}$$

which shows the effects of eating raw or incompletely cooked pork, being around infected cats, and handling dirt that could contain infective oocysts. The prevalence in pigs, in turn, is dependent on the density of infected cats.

We assume that the percent of raw/undercooked food eaten by humans is 1%. We also assume the infection rate four times higher for males than females. The resulting human growth and infection model is given in Figure 9.1.

We assume the initial rat population is 200, with a 10% birth rate (Figure 9.2). The initial value for the number of infected rats is based on a mean prevalence of 10%. The rat infection rate is 0.008 rats per year. Infected rats die at a rate that is dependent on the cat population, with the infected rats dying at rate that is 20% higher.

We assume an initial cat population of 200 with a 10% birth rate. The cat infection rate is dependent on both the number of infected cats and the number of infected rats.

The relationship for the spread among cats is based on the law of mass action so that the infection rate equals 1% of the product of infected and healthy cats plus an additive effect from the presence of infected rats (0.01*Inf rat density). The cats can then either become infected or become vaccinated. Infected cats stay contagious for 2 years, after which they become immune, a process which is represented by a conveyor. The infected cats die at a rate of 10%, while healthy cats die at a rate that is dependent on the total cat population. The cat population and infection process model is shown in Figure 9.3.

The small auxiliary models can be found in the model file.

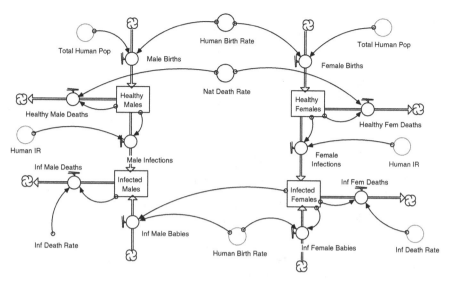

Fig. 9.1

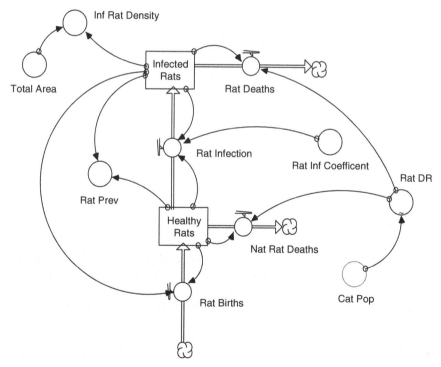

Fig. 9.2

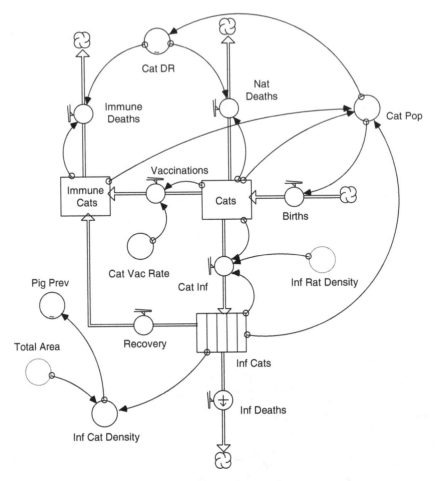

Fig. 9.3

9.3 Results

Figure 9.4 shows the numbers of healthy and infected people, as well as the prevalence of the disease in the humans over 25 years, with no vaccination of cats.

Figure 9.5 shows the effect of a 50% cat vaccination rate on the human prevalence
compared to no vaccinations. Vaccinating the cats does have a significant impact in
reducing the prevalence.

Figure 9.6 shows a sensitivity analysis of different cat vaccination rates of 40%,
50%, 60%, 70%, and 80%. A significant decrease in prevalence occurs between
40% and 50% vaccination rates.

Further runs of the model show that dirt handling had only a slight impact on
the disease in humans but that the effect of thorough meat cooking is extremely

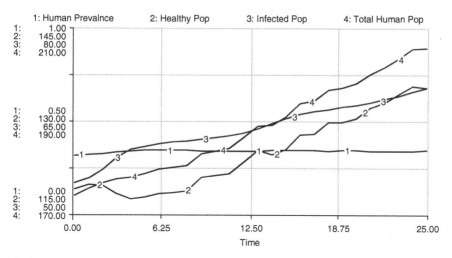

Fig. 9.4

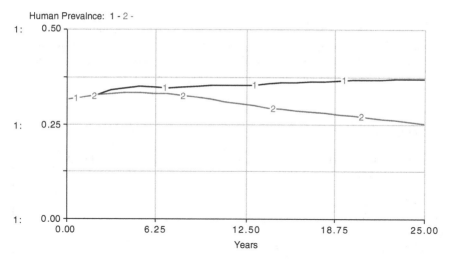

Fig. 9.5

important. In summary, both vaccinating cats at a rate greater than 52% and decreasing the amount of raw/undercooked meat that is eaten would significantly decrease the human prevalence of *T. gondii*.

9.4 Questions and Tasks

1. Include age differences in infection rates for humans in the model. What are the likely results, and how are the general conclusions in the earlier model affected?

Human Prevalnce: 1 - 2 - 3 - 4 - 5 -

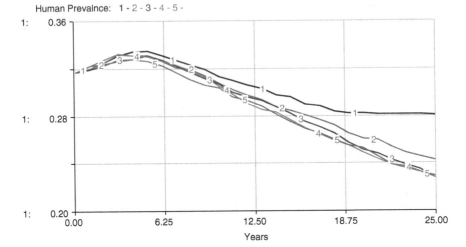

Fig. 9.6

2. Exert an exogenous shock onto the model by removing or adding a significant number of cats from the system at some random point in time. How are the results different? Are the results qualitatively different (aside from their opposite impacts) for additions and subtractions of cats? Explain your answer.

TOXOPLASMOSIS

Cats(t) = Cats(t − dt) + (Births − Vaccinations − Cat_Inf − Nat_Deaths) * dt
INIT Cats = 150

INFLOWS:
Births = Cat_Pop*0.1
OUTFLOWS:
Vaccinations = Cats*Cat_Vac_Rate
Cat_Inf = Cats*0.01*Inf_Cats + Inf_Rat_Density*0.01
Nat_Deaths = Cats*Cat_DR
Healthy_Females(t) = Healthy_Females(t − dt) + (Female_Births −
Female_Infections − Healthy_Fem_Deaths) * dt
INIT Healthy_Females = 27

INFLOWS:
Female_Births = Total_Human_Pop*Human_Birth_Rate/2
OUTFLOWS:
Female_Infections = Healthy_Females*Human_IR
Healthy_Fem_Deaths = Healthy_Females*Nat_Death_Rate
Healthy_Males(t) = Healthy_Males(t − dt) + (Male_Births − Male_Infections −

Healthy_Male_Deaths) * dt
INIT Healthy_Males = 92

INFLOWS:
Male_Births = Human_Birth_Rate*Total_Human_Pop/2
OUTFLOWS:
Male_Infections = Healthy_Males*Human_IR*4
Healthy_Male_Deaths = Healthy_Males*Nat_Death_Rate
Healthy_Rats(t) = Healthy_Rats(t − dt) + (Rat_Births − Rat_Infection −
Nat_Rat_Deaths) * dt
INIT Healthy_Rats = 200

INFLOWS:
Rat_Births = 0.1*(Healthy_Rats + Infected_Rats)
OUTFLOWS:
Rat_Infection = Rat_Inf_Coefficent*Healthy_Rats*Infected_Rats
Nat_Rat_Deaths = Healthy_Rats*Rat_DR
Immune_Cats(t) = Immune_Cats(t − dt) + (Vaccinations + Recovery −
Immune_Deaths) * dt
INIT Immune_Cats = 0

INFLOWS:
Vaccinations = Cats*Cat_Vac_Rate
Recovery = CONVEYOR OUTFLOW
OUTFLOWS:
Immune_Deaths = Immune_Cats*Cat_DR
Infected_Females(t) = Infected_Females(t − dt) + (Female_Infections +
Inf_Female_Babies − Inf_Fem_Deaths) * dt
INIT Infected_Females = 13

INFLOWS:
Female_Infections = Healthy_Females*Human_IR
Inf_Female_Babies = Human_Birth_Rate*(Infected_Females*.1)/2
OUTFLOWS:
Inf_Fem_Deaths = Infected_Females*Inf_Death_Rate
Infected_Males(t) = Infected_Males(t − dt) + (Male_Infections +
Inf_Male_Babies − Inf_Male_Deaths) * dt
INIT Infected_Males = 42

INFLOWS:
Male_Infections = Healthy_Males*Human_IR*4
Inf_Male_Babies = Infected_Females*.1*Human_Birth_Rate/2
OUTFLOWS:
Inf_Male_Deaths = Infected_Males*Inf_Death_Rate
Infected_Rats(t) = Infected_Rats(t − dt) + (Rat_Infection − Rat_Deaths) * dt
INIT Infected_Rats = 20

INFLOWS:
Rat_Infection = Rat_Inf_Coefficent*Healthy_Rats*Infected_Rats
OUTFLOWS:
Rat_Deaths = Infected_Rats*Rat_DR*1.2
Inf_Cats(t) = Inf_Cats(t − dt) + (Cat_Inf − Recovery − Inf_Deaths) * dt
INIT Inf_Cats = 50
   TRANSIT TIME = 2
   INFLOW LIMIT = ∞
   CAPACITY = ∞

INFLOWS:
Cat_Inf = Cats*0.01*Inf_Cats + Inf_Rat_Density*0.01
OUTFLOWS:
Recovery = CONVEYOR OUTFLOW
Inf_Deaths = LEAKAGE OUTFLOW
   LEAKAGE FRACTION = 0.10
   NO-LEAK ZONE = 0
Cat_Pop = Cats + Immune_Cats + Inf_Cats
Cat_Vac_Rate = {0.52}0
Dirt_Handling = 0.1
Healthy_Pop = Healthy_Females + Healthy_Males
Human_Birth_Rate = NORMAL(16/1000, 0.005)
Human_IR = (Percent_Raw_Food*Pig_Prev*1) + (Inf_Cat_Density*0.003) +
(Dirt_Handling*0.005)
Human_Prevalence = Infected_Pop/Total_Human_Pop
Infected_Pop = Infected_Females + Infected_Males
Inf_Cat_Density = Inf_Cats/Total_Area
Inf_Death_Rate = 0.02
Inf_Rat_Density = Infected_Rats/Total_Area
Nat_Death_Rate = NORMAL(1700/200000, 0.0005)
Percent_Raw_Food = 0.01
Rat_Inf_Coefficent = 0.0002
Rat_Prev = Infected_Rats/(Infected_Rats+Healthy_Rats)
Total_Area = 100
Total_Human_Pop = Healthy_Females + Healthy_Males+Infected_Females +
Infected_Males
Cat_DR = GRAPH(Cat_Pop)
(0.00, 0.0022), (50.0, 0.017), (100, 0.03), (150, 0.0382), (200, 0.051), (250,
0.069), (300, 0.0818), (350, 0.0968), (400, 0.113), (450, 0.128), (500, 0.148)
Pig_Prev = GRAPH(Inf_Cat_Density)
(0.00, 0.01), (0.05, 0.065), (0.1, 0.245), (0.15, 0.29), (0.2, 0.32), (0.25, 0.265),
(0.3, 0.285), (0.35, 0.27), (0.4, 0.29), (0.45, 0.265), (0.5, 0.275)
Rat_DR = GRAPH(Cat_Pop)
(0.00, 0.002), (50.0, 0.002), (100, 0.004), (150, 0.004), (200, 0.044), (250,
0.065), (300, 0.077), (350, 0.09), (400, 0.096), (450, 0.105), (500, 0.105)

Chapter 10
The Zebra Mussel[1]

10.1 Introduction

The zebra mussel (*Dreissena polymorpha*) is a small bivalve mollusk native to
Europe. The mussel was first observed in North American lakes only recently. Initial
colonization may have occurred in 1986, probably from larvae discharged in ballast
water. The zebra mussel is a potentially serious pest. In high densities, it presents
major problems for both human-made structures and for the ecology of infested bod-
ies of water. One of the most harmful impacts is the colonization of intake cribs and
pipes serving water treatment plants, power generating stations, and industries. At-
tracted to swift-moving water carrying large amounts of nutrients, mussels quickly
colonize and block these intake pipes. As an efficient feeder, the zebra mussel is
capable of removing large amounts of seston from the water. A benthic (bottom
dwelling) organism, the zebra mussel effectively removes nutrients from the water
column and deposits them on the bottom of the lake, river, or estuary in which it
lives. In addition to diverting primary productivity from the plankton to the benthos,
zebra mussels may also cover substrates used by other organisms and foul sedentary
benthic organisms.

10.2 Model Development

There is concern about the long-term effects of shifting large amounts of organic
matter from the pelagic to benthic zones. Effects of these dramatic changes in water
clarity and energy distribution on invertebrate, aquatic plant, and fish communities
have yet to be determined. Therefore, let us develop a model to investigate the potential impact that the introduction of zebra mussels will have on a small lake in
Northern America and forecast the potential growth of the zebra mussel in some

---

[1] This chapter is based on a project developed by Julie Sweitzer and Frederic Pieper for one of our
classes on dynamic modeling. We thank them for their contribution.

B. Hannon and M. Ruth, *Dynamic Modeling of Diseases and Pests*,
Modeling Dynamic Systems,
© Springer Science + Business Media LLC 2009

small lake. The questions to be answered by the model are: If zebra mussels are introduced into the lake, how will the population grow over time? What are the critical parameters that influence the size of the population? How significant an impact will the equilibrium mussel population have on the lake's ecosystem?

In order to enhance the understanding of the model, we split it up into three parts, or modules, that are interrelated and together comprise the critical dynamic aspects of the questions being addressed. Partitioning a model into individual, easily comprehensible parts is always helpful in making the model structure transparent. In the model of the zebra mussel, the growth module captures the growth processes of the zebra mussel population. The sustainability module determines the long-term, sustainable level of the population by incorporating some key physical parameters. The filtration module yields a relative indication of the impact of the population on the lake by computing the frequency with which the total population filters the entire water volume of the lake. An explanation of the structure of each of these modules and the data and assumptions that were used to construct them follows.

The growth module in Figure 10.1 is made up of four population cohorts (one juvenile and three adult). The average life span of a zebra mussel appears to vary between populations in different areas. We chose a life span of four years, which seemed most appropriate for the analysis. Zebra mussel populations typically reproduce once a year (usually sometime during the summer) according to a mass synchronous spawning behavior. Therefore, each cohort represents one generation of mussels. The juvenile cohort represents all those mussels that have successfully attached to some suitable substrate and have grown to a certain average size by the end of the first year. These individuals are not yet sexually mature. The three adult cohorts are the populations of each generation that has reached sexual maturity and survived to the end of years two, three, and four.

The number of individual mussels that enter the juvenile cohort in a certain year is a function of the number of adult female zebra mussels and their fecundities. Although fecundities are high (30,000–40,000 larvae per female), larval mortality rates are fairly high as well. The survival rate (.008) represents the percent of larvae that attach to a substrate and grow to a certain size by the end of the year.

The survival potential is then the number of mussels that reach the juvenile cohort under normal circumstances (no limiting conditions). However, the actual number of mussels that survive the first year depends on whether or not the overall population is close to or above the sustainable population of the lake. The sustainable population is determined in the sustainability module described in detail below. So, the actual number that survive is calculated as follows: if the total population is less than the sustainable population in any year, then SURVIVE equals the smaller of the survival potential, SURVIVE POT, and the difference between SUSTAINABLE POPULATION and TOTAL POPULATION. If the total population is greater than or equal to the sustainable population, then SURVIVE equals SURVIVE POT times a factor that is less than one and decreases exponentially as total population gets larger. The result is that the more the total population exceeds the sustainable population, the smaller the number of juveniles that survive during that year.

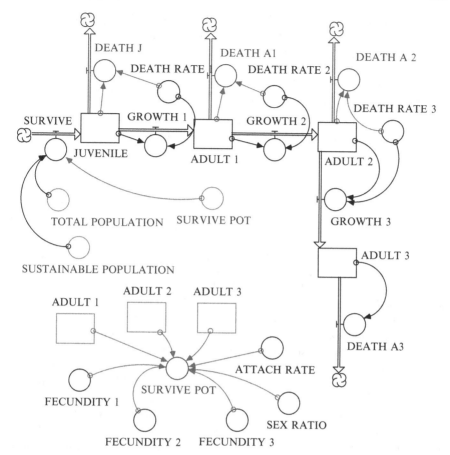

Fig. 10.1

At first glance, it appears that the sustainable population will not be exceeded. However, due to the way in which the populations move between cohorts, the total population does, in fact, overshoot the sustainable level in certain years. Adult populations die at certain rates between years; the surviving population grows to a new average size and enters the next cohort. At the end of the fourth year, it is assumed that the entire generation dies out. The sustainability module of the model is shown in Figure 10.2.

The sustainability module determines the equilibrium population level that is expected to be reached over the long term. The key parameter that determines this level is the availability of suitable substrate material upon which the mussels can attach. Zebra mussels require hard substrates and cannot live in muddy conditions. Other potentially limiting factors, such as calcium deficiency or extremes in water temperature or nutrient availability, are not considered here.

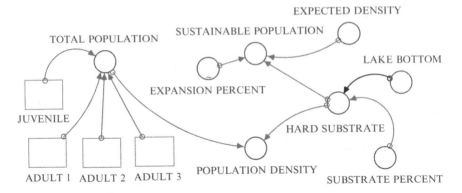

Fig. 10.2

Available hard substrate is calculated by multiplying the total area of the lake bottom by the percent of the lake bottom that is hard. This percentage can only be estimated, and its initial value is given. To determine the sustainable population at any point in time, the average density and distribution of the population must also be known. Observed density figures vary widely in the literature. Therefore, we have chosen a range over which we run our model. Our expected density figure is taken to be that density which is sustainable over the long term.

The expansion percent parameter shows the percent of the lake that has been colonized over time and is represented in graphical form. We are assuming that infestation occurs at a specific point (as from an infested bait bucket dumped overboard) and spreads from there. Zebra mussel larvae are distributed via lake currents as well as boat traffic. We assume that in the first month, 20% of the lake is infested with the mussel. The expansion percentage increases drastically during the first few months and reaches a maximum of 100% by the end of the first year. The percentage expansion remains at that level for the relevant future. For simplicity, we assume the values shown in Figure 10.3.

Try experimenting with the expansion rate in alternative runs of the model. For example, assume that the expansion percentage is reduced to 20% at the beginning of each year, and increases toward 100% over the course of the year. This processes is then repeated each year. Toward that end, the EXPANSION PERCENT can be modeled with the built-in function "MOD." MOD(TIME,12)+1 converts simulation time into months, starting at 1. After 12 months of simulation time, MOD(TIME,12)+1 will reset itself to 1.

In the filtration module of Figure 10.4, LAKE TURNOVER is calculated. This serves as a relative indication of how great an impact the population will have on the current lake ecosystem. LAKE TURNOVER is the number of times per day that the total mussel population filters the total volume of water in the lake. The filtration rate of an individual mussel is a function of its shell length and is named here FILTER1, FILTER2, and so forth, for the respective age classes 1, 2, and so on. An initial shell length, LENGTH1, LENGTH2, and so forth, is assumed for the

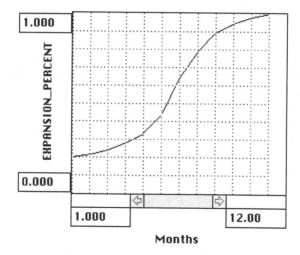

Fig. 10.3

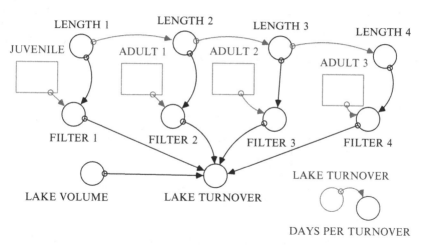

Fig. 10.4

juvenile population, and shell length grows each year as a function of the previous year's length.

The amount of water filtered by each cohort is calculated by multiplying the individual filtration rates times the population of each cohort. Daily lake turnover is the sum of the daily filtration volumes of each cohort divided by the total lake volume. In addition, the number of days per complete lake turnover is calculated by dividing the daily lake turnover into one complete turn.

10.3 Model Results

Now the model is complete and can be used to develop a base case model of zebra mussel growth in our lake. The base case model uses average values for model parameters published in the literature[2]. You will find that data in the model equations at the end of the chapter.

The results of the base case model can be compared with alternative scenarios to determine the impact and sensitivity of various parameters on the growth patterns. Such a procedure is important if some data are unknown or unavailable, or if there is uncertainty surrounding parameter estimates.

For a sensitivity analysis, vary the value for SUBSTRATE PERCENT between 50% and 90% and the death rates for the first three age classes within the intervals listed in Table 10.1:

Change only one parameter value at a time.

The base case growth, density, and lake turnover pathways are shown in Figure 10.5. We have assumed that the initial invading population consists of 100 juveniles. In the early years, population grows exponentially and reaches a total population peak of $8.08*10^8$ mussels in year thirteen. The population then oscillates with decreasing extremes toward a steady state population of $7.79*10^8$. Population density follows the same pattern as total population growth. Density peaks at $834\,mussels/m^2$ and stabilizes at $772\,mussels/m^2$.

Table 10.1

	Minimum	Maximum
DEATH RATE 1	0.06	0.12
DEATH RATE 2	0.09	0.15
DEATH RATE 3	0.18	0.18

---

[2] See, for example: Bij de Vaate, A. 1991. Distribution and Aspects of Population Dynamics of the Zebra Mussel, *Dreissena polymorpha* (Pallas, 1771), in the Lake Ijsselmeer Area (The Netherlands). *Oecologia* Vol. 86, pp. 40–50; Griffiths, R.W., W.P. Kovalak, and D.W. Schloesser. 1989. The Zebra Mussel, *Dreissena polymorpha*, in North America: Impact on Raw Water Users, in *Proceedings: EPRI Service Water System Reliability Improvement Seminar*. Electric Power Research Institute, Palo Alto, CA, pp. 11–27; Griffiths, R.W., D.W. Schloesser, J.H. Leach, and W.P. Kovalak. 1991. Distribution and Dispersal of the Zebra Mussel (*Dreissena polymorpha*) in the Great Lakes Region, *Canadian Journal of Fisheries and Aquatic Science*, Vol. 48, pp. 1381–1388; Haag, W.R. and D.W. Garton. 1992. Synchronous Spawning in a Recently Established Population of the Zebra Mussel, *Dreissena polymorpha*, in Western Lake Erie, USA., *Hydrobiologia*, Vol. 234, pp. 103–110; Kryger, J. and H.U. Riisgard. 1988. Filtration Rate Capacities in 6 Species of European Freshwater Bivalves., *Oecologia*, Vol. 77, pp. 34–38; Mackie, G. 1991. Biology of the Exotic Zebra Mussel, *Dreissena polymorpha*, in Relation to Native Bivalves and its Potential Impact in Lake St. Clair, *Hydrobiologia*, Vol. 219, pp. 251–268; Strayer, D.L. 1991. Projected Distribution of the Zebra Mussel, *Dreissena polymorpha*, in North America, *Canadian Journal of Fisheries and Aquatic Science*, Vol. 48, pp. 1389–1395.

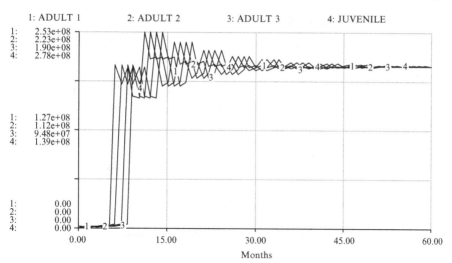

Fig. 10.5

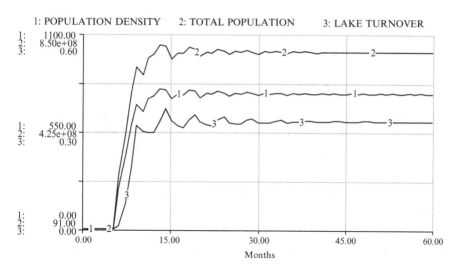

Fig. 10.6

Similarly, lake turnover (Figure 10.6) closely follows the population pattern since lake turnover is directly related to the number of zebra mussels in the lake, but it is slightly lagged. Lake turnover peaks in year fourteen at 0.37 (total lake turnover per day) and stabilizes at 0.33. This variable is more intuitive if presented as the number of days required for the mussel population to filter the entire volume of the lake. In the peak of the base case, year fourteen, the mussel population filters the volume of the lake in 2.69 days. This rate changes as population drops and stabilizes at 2.99 days/turnover.

10.4 Questions and Tasks

1. Explore the impacts of extreme weather conditions (e.g. a period of extreme summer temperatures or drought) on zebra mussel dynamics.
2. How would you introduce the impacts of a chemical eradication program on zebra mussels? Can you get rid of all the mussels from the lake? If so, for how long?

ZEBRA MUSSEL MODEL

ADULT_1(t) = ADULT_1(t − dt) + (GROWTH_1 − GROWTH_2 − DEATH_A1) * dt
INIT ADULT_1 = 0 {Number of Individuals}
INFLOWS:
GROWTH_1 = (1 − DEATH_RATE_1)*JUVENILE {Individuals per Month}
OUTFLOWS:
GROWTH_2 = (1 − DEATH_RATE_2)*ADULT_1 {Individuals per Month}
DEATH_A1 = DEATH_RATE_2*ADULT_1 {Individuals per Month}

ADULT_2(t) = ADULT_2(t − dt) + (GROWTH_2 − GROWTH_3 − DEATH_A_2) * dt
INIT ADULT_2 = 0 {Number of Individuals}
INFLOWS:
GROWTH_2 = (1 − DEATH_RATE_2)*ADULT_1 {Individuals per Month}
OUTFLOWS:
GROWTH_3 = (1 − DEATH_RATE_3)*ADULT_2 {Individuals per Month}
DEATH_A_2 = DEATH_RATE_3*ADULT_2 {Individuals per Month}
ADULT_3(t) = ADULT_3(t − dt) + (GROWTH_3 − DEATH_A3) * dt
INIT ADULT_3 = 0 {Number of Individuals}
INFLOWS:
GROWTH_3 = (1 − DEATH_RATE_3)*ADULT_2 {Individuals per Month}
OUTFLOWS:
DEATH_A3 = 1.0*ADULT_3 {Individuals per Month}

JUVENILE(t) = JUVENILE(t − dt) + (SURVIVE − GROWTH_1 − DEATH_J) * dt
INIT JUVENILE = 100 {Number of Individuals}
INFLOWS:
SURVIVE = IF (TOTAL_POPULATION<SUSTAINABLE_POPULATION) THEN MIN(SURVIVE_POT,SUSTAINABLE_POPULATION-TOTAL_POPULATION)
ELSE ((1 − (TOTAL_POPULATION-SUSTAINABLE_POPULATION)/TOTAL_POPULATION)^2)*SURVIVE_POT {Number of Individuals per Month}
OUTFLOWS:
GROWTH_1 = (1 − DEATH_RATE_1)*JUVENILE {Individuals per Month}
DEATH_J = DEATH_RATE_1*JUVENILE {Individuals per Month}

ATTACH_RATE = .8*.01 {20% larval mortality in veliger stage, 99% mortality post-veliger; Bij de Vaate 1991 p.46; Percent of Larvae Which Successfully Attach to Suitable Substrate and Survive at Least to the End of the Year}

DAYS_PER_TURNOVER = 1/LAKE_TURNOVER {number of days for complete lake turnover}
DEATH_RATE_1 = .09
{Bij de Vaate, 1991,p.10}
DEATH_RATE_2 = .12
DEATH_RATE_3 = .15
EXPECTED_DENSITY = 1000
{sq m-2; Strayer; Ch 43}
FECUNDITY_1 = 30000 {Mackie, 1991; Larvae per Female}
FECUNDITY_2 = 35000 {Larvae per Female}
FECUNDITY_3 = 40000 {Mackie, 1991; Larvae per Female}
FILTER_1 = JUVENILE*(6.82*(1.54E − 5*LENGTH_1^2.42)^.88)
{L/hr; Kryger and Rilsgard, 1988}
FILTER_2 = ADULT_1*(6.82*(1.54E − 5*LENGTH_2^2.42)^.88)
{L/hr; Kryger and Rilsgard, 1988}
FILTER_3 = ADULT_2*(6.82*(1.54E − 5*LENGTH_3^2.42)^.88)
{L/hr; Kryger and Rilsgard, 1988}
FILTER_4 = ADULT_3*(6.82*(1.54E − 5*LENGTH_4^2.42)^.88)
{L/hr; Kryger and Rilsgard, 1988}
HARD_SUBSTRATE = LAKE_BOTTOM*SUBSTRATE_PERCENT
{sq. m}
LAKE_BOTTOM = (1.3*1E+6+0.142*1E+6)
LAKE_TURNOVER =
(FILTER_1+FILTER_2+FILTER_3+FILTER_4)*24/LAKE_VOLUME
{total lake turnover per day}
LAKE_VOLUME = 8524472*1000 {liters(cubic meters * 1000L/cu.M); BLA}
LENGTH_1 = 5 {mm}
LENGTH_2 = LENGTH_1+(.006*(LENGTH_1^2)-.56*LENGTH_1+12.1)
{Bij de Vatte, 1991; mm}
LENGTH_3 = LENGTH_2+(.006*(LENGTH_2^2)−.56*LENGTH_2+12.1)
{mm}
LENGTH_4 = LENGTH_3+(.006*(LENGTH_3^2)−.56*LENGTH_3+12.1)
{mm}
POPULATION_DENSITY = TOTAL_POPULATION/HARD_SUBSTRATE
SEX_RATIO = .6
{Mackie, 1991,p.255}

SUBSTRATE_PERCENT = .70
SURVIVE_POT =
ATTACH_RATE*(SEX_RATIO*(FECUNDITY_1*ADULT_1+
FECUNDITY_2*ADULT_2+FECUNDITY_3*ADULT_3)) {Number of Individuals}

SUSTAINABLE_POPULATION = EXPANSION_PERCENT*
HARD_SUBSTRATE*EXPECTED_DENSITY {Number of Individuals}
TOTAL_POPULATION = JUVENILE+ADULT_1+ADULT_2+ADULT_3
{Number of Individuals}
EXPANSION_PERCENT = GRAPH(TIME)
(1.00, 0.2), (2.00, 0.215), (3.00, 0.24), (4.00, 0.28), (5.00, 0.33), (6.00, 0.435),
(7.00, 0.64), (8.00, 0.78), (9.00, 0.89), (10.0, 0.945), (11.0, 0.98), (12.0, 1.00)

Chapter 11
Biological Control of Pestilence

Throughout this book we have hinted at—or explicitly modeled—strategies that interfere with the dynamics of pests and diseases, such as using repellents in the malaria model or vaccination in the case of chicken pox. In this chapter, we focus on such interferences and concentrate on biological methods to control pests. The subsequent chapter then explores the effects of disease resistance.

11.1 Herbivory and Algae

11.1.1 Herbivore-Algae Predator-Prey Model

The first of our biological pestilence control models uses a simple predator–prey model to show that even without migration, the system can exhibit a wide range of responses. Assume that the prey are algae in a pond on which an herbivore grazes. The data for this problem have been invented. (Normally, input data, parameters, and initial conditions would be determined by experiment.)

The model consists of two main parts: one is for the change in the algae population, and one is for the herbivore. The growth rate is a function of the algal density, ALGAE. This function is monotonic and declining (Figure 11.1). Algal growth is calculated as the product of the density and the growth rate.

The algae density is reduced through consumption by the herbivore. The consumption per head is a nonlinear function of the algal density: the greater the density, the higher the consumption per head (Figure 11.2). The consumption rate is simply the product of the number of herbivore and the consumption per head.

The herbivore death rate is determined by their average life span, which is a nonlinear function of the consumption per head: the higher the consumption per head, the longer the life span, within limits (Figure 11.3). Indirectly, the denser the algae, the lower the herbivore death rate.

B. Hannon and M. Ruth, *Dynamic Modeling of Diseases and Pests,*
Modeling Dynamic Systems,
© Springer Science+Business Media LLC 2009

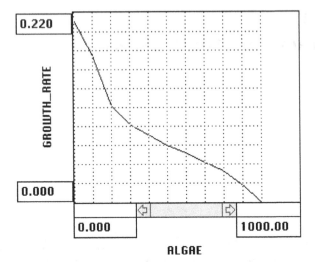

Fig. 11.1

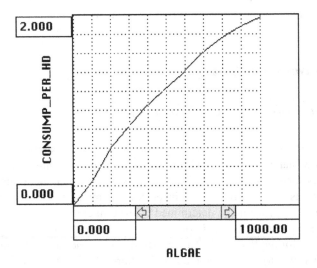

Fig. 11.2

The herbivore growth rate is a product of the herbivore stock and the fractional herbivore growth rate, FCN HERB GROW (Figure 11.4). To increase realism of the model, we make FCN HERB GROW a function of the algae density in the previous time period. This is done by producing an additional stock called ALGAE DELAY. In general, it makes sense to represent herbivore behavior in this way. Herbivore gestation time reflects the origin of this lagged behavior.

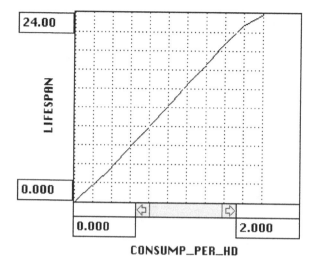

Fig. 11.3

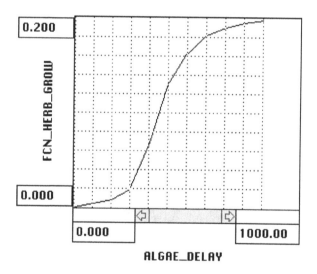

Fig. 11.4

The combination of the assumptions shown in Figures 11.1 through 11.4, together with the basic alga and herbivore population model, is shown in Figure 11.5.

Figure 11.6 shows the wide swings in algal density and herbivore population over time. Figure 11.7 presents a plot of algal density against the herbivore population and exhibits the limit cycle resulting from this particular choice of the variables.

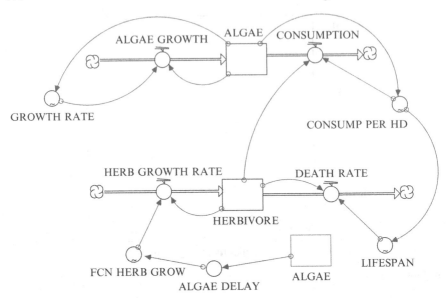

Fig. 11.5

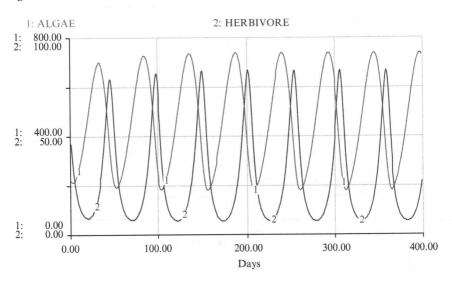

Fig. 11.6

11.1.2 *Questions and Tasks*

1. Can you make the herbivore of this model crash and not reemerge?
2. Try to maximize the herbivore population. Can you do this by adjusting only
   the variable FCN HERB GROW, without changing the maximum and minimum
   rates?

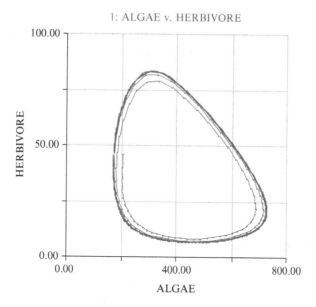

Fig. 11.7

HERBIVORE-ALGAE MODEL

ALGAE(t) = ALGAE(t − dt) + (ALGAE_GROWTH − CONSUMPTION) * dt
INIT ALGAE = 210 {Algae per Area}
INFLOWS:
ALGAE_GROWTH = ALGAE * GROWTH_RATE {Algae per Area per Time Period}
OUTFLOWS:
CONSUMPTION = HERBIVORE * CONSUMP_PER_HD {Algae per Area per Time Period}

HERBIVORE(t) = HERBIVORE(t − dt) + (HERB_GROWTH_RATE − DEATH_RATE) * dt
INIT HERBIVORE = 45 {Individuals}
INFLOWS:
HERB_GROWTH_RATE = HERBIVORE * FCN_HERB_GROW {Individuals per Time Period}
OUTFLOWS:
DEATH_RATE = HERBIVORE/LIFESPAN {Individuals per Time Period}

ALGAE_DELAY = DELAY(ALGAE,2) {Individuals}
CONSUMP_PER_HD = GRAPH(ALGAE)
(0.00, 0.00), (100, 0.25), (200, 0.6), (300, 0.83), (400, 1.06), (500, 1.24), (600, 1.41), (700, 1.61), (800, 1.77), (900, 1.89), (1000, 1.98)

FCN_HERB_GROW = GRAPH(ALGAE_DELAY)
(0.00, 0.00), (100, 0.0035), (200, 0.0075), (300, 0.019), (400, 0.065), (500, 0.13),
(600, 0.163), (700, 0.181), (800, 0.19), (900, 0.195), (1000, 0.198)
GROWTH_RATE = GRAPH(ALGAE)
(0.00, 0.21), (100, 0.168), (200, 0.112), (300, 0.0902), (400, 0.0781), (500,
0.066), (600, 0.0572), (700, 0.0462), (800, 0.0363), (900, 0.0198), (1000, 0.00)
LIFESPAN = GRAPH(CONSUMP_PER_HD)
(0.00, 0.00), (0.2, 2.16), (0.4, 4.32), (0.6, 6.96), (0.8, 9.48), (1.00, 12.1), (1.20,
14.9), (1.40, 17.3), (1.60, 20.2), (1.80, 22.6), (2.00, 23.8)

11.2 Bluegill Population Management

11.2.1 Bluegill Dynamics

The small game fish population in human-made reservoirs is a continual problem.
Without a balanced set of predators and their prey and without natural conditions in
the reservoirs, fish stocked in the reservoirs tend to either disappear or to produce a
very large number of undersized fish. The demise of fish has two main causes: death
due to natural causes (predation, disease) and fishing. Fishing pressure is the main
reason for the disappearance of the larger members of the game species. The management of fishing provides a method to produce a more balanced fish population,
and here we demonstrate how to model such regulations for the bluegill in Illinois
reservoirs.

We are interested in modeling[1] the effects of various factors influencing size
structure of a bluegill population over a 50-year period. We wish to simulate population dynamics on a temporal scale in an effort to provide fishery managers with
information necessary to manage bluegill populations in Illinois reservoirs. This
model could serve as a tool to test proposed management regulations on a given
reservoir to provide insight into the changes resulting from the proposed regulations.

The model is developed to answer the following questions:

1. How do bluegill populations change through time when no fishing occurs?
2. How do bluegill populations behave through time when fishing is allowed and
   there are no fishing regulations?
3. How does implementation of management regulations, such as a creel limit and
   a size limit, affect the bluegill population?
4. Are there differences in the way in which male and female bluegill populations
   respond to such management scenarios?
5. Is the overall bluegill population sensitive to changes in the female bluegill population?

---

[1] We thank Brian Herwig and Derek Aday for their contributions to this model.

We have created a model that explores factors influencing the growth, removal, and size structure of bluegill populations. These factors include density dependent growth, natural mortality (a variable that accounts for losses due to predation, disease, winterkill, and starvation), and finally, losses due to removal via recreational angling (the variable of interest in this model). Often these processes are size and sex specific. To account for this, we divided the bluegill population into seven 30-mm size categories from a 0- to 30-mm cohort up to a 180- to 210-plus-mm cohort. To account for difference between sexes, we split the population into a male component and a female component to give our model further resolution (Figure 11.8). We added a section to account for egg production and the resulting fry production of the population. Following the fry stage of the model, we included a component to determine the sex and life history strategy of the fry. The sex ratio of the population was assumed to be 50:50. The life history component was determined through a submodel, which bases life history of male fry on the proportion of adult males in the population (the greater the density of adult males, the greater the proportion of male fry that become "sneakers"). Sneakers are male bluegills that adopt a cuckoldry life history strategy. The numbers used throughout the model were averages from data collected from numerous fish populations around the Midwest. Numbers for parameters that were not available from empirical data were estimated from discussions with fishery biologists in the Illinois Natural History Survey in 1997.

To address the questions listed above we ran the model three ways:

1. With no fishing mortality and with just growth and natural mortality occurring, to establish a baseline with which to test the effects of management regulations on the population.
2. With growth and natural mortality plus mortality due to fishing, where fishing is unregulated and set by the number of anglers, hours fished, and the catchability of fish in the cohort.
3. With growth and natural mortality plus mortality due to fishing, but where fishing is regulated such that a creel is set for selected fishable cohorts, establishing an effective size limit and maximum number of fish that can be removed from the population per year.

In the case where there is no fishing mortality, similar patterns emerge for similar size classes for both males and females (Figures 11.9 and 11.10). There are a greater number of individuals (150,000 to 200,000) in the smaller size classes (0 to 30 mm and 30 to 60 mm), and a smaller number of fish (70,000 to 90,000) in the larger size classes (>90 mm). This represents a typical size distribution found in many lakes. After some initial fluctuation, the selected cohorts stabilize around the ranges stated earlier, but increase slightly through time.

Note that in these graphs and in all the following ones, a period of approximate 6 years is required for the population to come to equilibrium. This is due to the initially stocked cohort of fry moving through all of the size classes before reaching equilibrium. Also note that the youngest size class (0 to 30 mm) is increasing most rapidly.

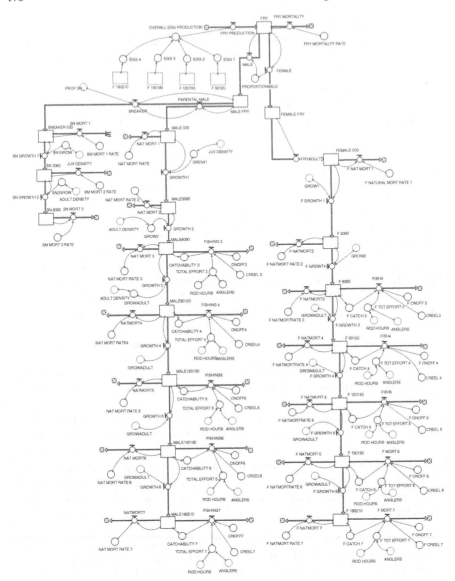

Fig. 11.8

11.2.2 Impacts of Fishing

When fishing mortality is imposed in the model, the fishable cohorts (60 to 90 mm through 180 to 210 mm), after some initial fluctuation, settle into equilibrium at population levels considerably lower than when fishing did not occur (55,000 to 63,000 with fishing compared to 70,000 to 90,000 without fishing). As before, more

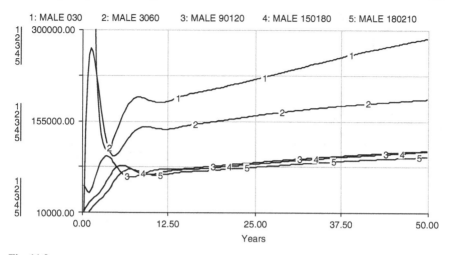

Fig. 11.9

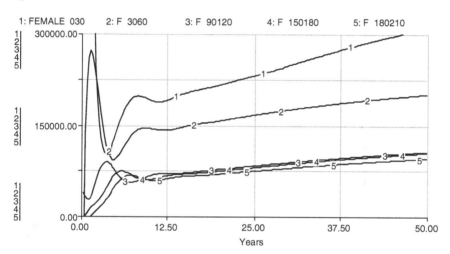

Fig. 11.10

fish are always in the smaller size classes. Note that the two largest cohorts are the lowest and are proportionately lower than those without fishing. These size classes are particular susceptible to fishing mortality because of their increased vulnerability and because anglers prefer to keep larger the larger fish (Figures 11.11 and 11.12).

Notice that the female population (Figure 11.12) is not as affected by fishing mortality as the male population (Figure 11.11). This is due to the fact that the catchability of females is not as high as males. Males generally remain higher in the water column and possess more aggressive behavioral characteristics that cause them to be more susceptible to fishing.

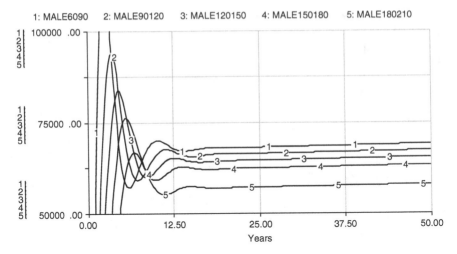

Fig. 11.11

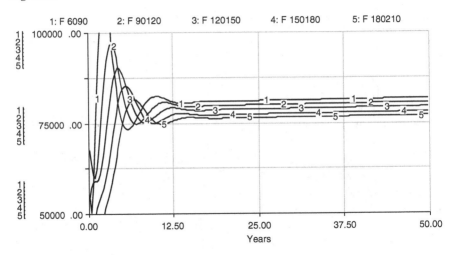

Fig. 11.12

To demonstrate the effect of a creel limit on the population of fish, we imposed a 100 fish limit on the following sizes of fish - 60 – 90 mm; 150 – 180 mm; 180 – 210 mm (the smallest and two largest fishable size classes). The result is that the entire fish population is affected (Figure 11.13).This is expected, because leaving adult fish (capable of reproduction) in the population results in more egg production and fertilization, and subsequently more fish in each size class. The numbers of fish actually return to levels similar to the case when there was no fishing mortality (70,000 – 90,000 fish).

In terms of sneaker males, an interesting result occurs. When a creel limit is imposed, the number of sneaker males increases (Figure 11.14). This is due to the

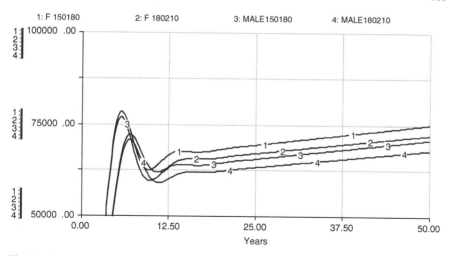

Fig. 11.13

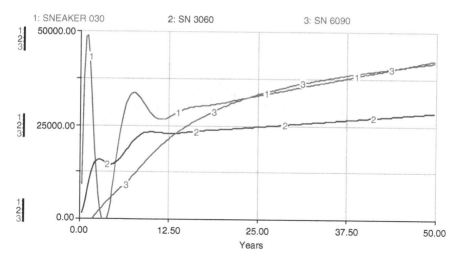

Fig. 11.14

fact that more large males remain in the population—the greater the population density of males, the greater the proportion of sneakers. When no creel limit is imposed, the proportion of sneakers drops, due to the decreased number of large males (Figure 11.15). This is an important finding, because the number of sneaker males in a population of bluegill is a concern to fishery managers. Large proportions of sneaker males in the population results in smaller sized fish and greater abundance of stunted fish as a result of evolutionary and genetic changes in the population.

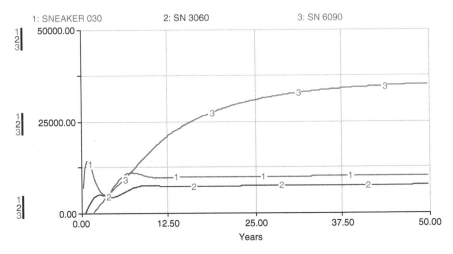

Fig. 11.15

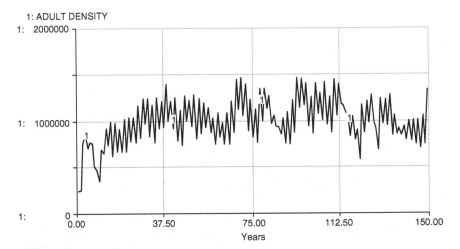

Fig. 11.16

11.2.3 Impacts of Disease

Disease was added to the model, and its results are shown in Figure 11.16. The impact of the disease is arbitrarily chosen to demonstrate the impact. If the adult density exceeded a limit $(750,000 * \text{random}(.8, 1.2)$ then the disease occurs, multiplying the natural mortality rate by a random increase between 1 and 3. With fishing regulations in place, the adult density (total number of fish in this fishery) fluctuates rather widely.

The disease decimates some of the cohorts but the population quickly recovers due to the hardiness of those in the reproductive age classes. The total adult density seems to fluctuate around a value of about 1 million, down from a disease-free adult density of about 1.2 million.

This model replicates the complex interaction between the sneaker male and adult male population, and demonstrates the trade-off between successfully managing for adult bluegill while trying to keep the sneaker population to a minimum. This model allows resource managers to test the effect of creel limits on the population of sneaker males before actually integrating the solutions in the management plan of a natural population. Could the model be made more instructive by splitting natural mortality into its components (i.e., winterkill, predation, disease, and so forth)? Although the precision of the model may increase due to accuracy of these measurements, would the complexity be increased to the point that the model is not useful? The reader is encouraged to expand the model in this way.

11.2.4 Questions and Tasks

1. a) Introduce a size limit on catch and observe its results.
   b) Can you find size restrictions (maximum and/or minimum allowable size of the catch) that stabilize the bluegill population for a given fishing pressure?
2. Assume the disease introduced in earlier discussion only affects juveniles. What are the effects on the size of the population as a whole, and on the average size of fish in the population?

BLUEGILL FISHERY MANAGEMENT

FEMALE_030(t) = FEMALE_030(t − dt) + (FRYADULT − F_GROWTH_1 − F_NAT_MORT_1) * dt
INIT FEMALE_030 = 1000000

INFLOWS:
FRYADULT = FEMALE_FRY
OUTFLOWS:
F_GROWTH_1 = FEMALE_030 * GROW1
F_NAT_MORT_1 = FEMALE_030 * F_NATURAL_MORT_RATE_1
FEMALE_FRY(t) = FEMALE_FRY(t − dt) + (FEMALE − FRYADULT) * dt
INIT FEMALE_FRY = 1000000

INFLOWS:
FEMALE = FRY * (1 − PROPORTIONMALE)
OUTFLOWS:
FRYADULT = FEMALE_FRY
FRY(t) = FRY(t − dt) + (FRY_PRODUCTION − FRY_MORTALITY −

FEMALE − MALE) * dt
INIT FRY = 200000

INFLOWS:
FRY_PRODUCTION = OVERALL_EGG_PRODUCTION * 0.000065
OUTFLOWS:
FRY_MORTALITY = FRY_MORTALITY_RATE * FRY
FEMALE = FRY * (1−PROPORTIONMALE)
MALE = FRY * PROPORTIONMALE
F_120150(t) = F_120150(t − dt) + (F_GROWTH_4 − FISH5 −
F_GROWTH_5 − F_NATMORT_5) * dt
INIT F_120150 = 30000

INFLOWS:
F_GROWTH_4 = F_90120 * GROWADULT
OUTFLOWS:
FISH5 = IF F_ONOFF_5 = 1 THEN (IF F_CATCH_5 * F_TOT_EFFORT_5 >
CREEL_5 THEN CREEL_5 ELSE F_CATCH_5 * F_TOT_EFFORT_5) ELSE IF
F_ONOFF_5 = 0 THEN F_CATCH_5 * F_TOT_EFFORT_5 ELSE 0
F_GROWTH_5 = F_120150 * GROWADULT
F_NATMORT_5 = F_NATMORTRATE_5 * F_120150
F_150180(t) = F_150180(t − dt) + (F_GROWTH_5 − F_MORT_6 −
F_GROWTH_6 − F_NATMORT_6) * dt
INIT F_150180 = 10000

INFLOWS:
F_GROWTH_5 = F_120150 * GROWADULT
OUTFLOWS:
F_MORT_6 = IF F_ONOFF_6 = 1 THEN (IF F_CATCH_6 * F_TOT_EFFORT_6
> CREEL_6 THEN CREEL_6 ELSE F_CATCH_6 * F_TOT_EFFORT_6) ELSE
IF F_ONOFF_6 = 0 THEN F_CATCH_6 * F_TOT_EFFORT_6 ELSE 0
F_GROWTH_6 = F_150180 * GROWADULT
F_NATMORT_6 = F_150180 * F_NATMORTRATE_6
F_180210(t) = F_180210(t − dt) + (F_GROWTH_6 − F_MORT_7 −
F_NATMORT_7) * dt
INIT F_180210 = 1000

INFLOWS:
F_GROWTH_6 = F_150180 * GROWADULT
OUTFLOWS:
F_MORT_7 = IF F_ONOFF_7 = 1 THEN (IF F_CATCH_7 * F_TOT_EFFORT_7
> F_CREEL_7 THEN F_CREEL_7 ELSE F_CATCH_7 * F_TOT_EFFORT_7)
ELSE IF F_ONOFF_7 = 0 THEN F_CATCH_7 * F_TOT_EFFORT_7 ELSE 0
F_NATMORT_7 = F_NATMORT_RATE_7 * F_180210

F_3060(t) = F_3060(t − dt) + (F_GROWTH_1 − F_GROWTH_2 − F_NATMORT2) * dt
INIT F_3060 = 10000

INFLOWS:
F_GROWTH_1 = FEMALE_030 * GROW1
OUTFLOWS:
F_GROWTH_2 = F_3060 * GROW2
F_NATMORT2 = F_3060 * F_NATMORT_RATE_2
F_6090(t) = F_6090(t − dt) + (F_GROWTH_2 − FISH3 − F_GROWTH_3 − F_NATMORT3) * dt
INIT F_6090 = 30000

INFLOWS:
F_GROWTH_2 = F_3060 * GROW2
OUTFLOWS:
FISH3 = IF F_ONOFF_3 = 1 THEN (IF F_CATCH_3 * F_TOT_EFFORT_3 > CREEL3 THEN CREEL3 ELSE F_CATCH_3 * F_TOT_EFFORT_3) ELSE IF F_ONOFF_3 = 0 THEN F_CATCH_3 * F_TOT_EFFORT_3 ELSE 0
F_GROWTH_3 = GROWADULT * F_6090
F_NATMORT3 = F_6090 * F_NATMORTRATE_3
F_90120(t) = F_90120(t − dt) + (F_GROWTH_3 − FISH4 − F_GROWTH_4 − F_NATMORT_4) * dt
INIT F_90120 = 50000

INFLOWS:
F_GROWTH_3 = GROWADULT * F_6090
OUTFLOWS:
FISH4 = IF F_ONOFF_4 = 1 THEN (IF F_CATCH_4 * F_TOT_EFFORT_4 > CREEL_4 THEN CREEL_4 ELSE F_CATCH_4 * F_TOT_EFFORT_4) ELSE IF F_ONOFF_4 = 0 THEN F_CATCH_4 * F_TOT_EFFORT_4 ELSE 0
F_GROWTH_4 = F_90120 * GROWADULT
F_NATMORT_4 = F_NATMORTRATE_4 * F_90120
MALE120150(t) = MALE120150(t − dt) + (GROWTH_4 − FISHING5 − GROWTH_5 − NATMORT5) * dt
INIT MALE120150 = 30000

INFLOWS:
GROWTH_4 = MALE90120 * GROWADULT
OUTFLOWS:
FISHING5 = IF ONOFF5 = 1 THEN (IF CATCHABILITY_5 * TOTAL_EFFORT_5 > CREEL5 THEN CREEL5 ELSE CATCHABILITY_5 * TOTAL_EFFORT_5) ELSE IF ONOFF5 = 0 THEN CATCHABILITY_5 * TOTAL_EFFORT_5 ELSE 0
GROWTH_5 = MALE120150 * GROWADULT
NATMORT5 = MALE120150 * NAT_MORT_RATE_5
MALE150180(t) = MALE150180(t − dt) + (GROWTH_5 − FISHING6 − GROWTH_6 − NAT_MORT6) * dt

INIT MALE150180 = 10000

INFLOWS:
GROWTH_5 = MALE120150 * GROWADULT
OUTFLOWS:
FISHING6 = IF ONOFF6 = 1 THEN (IF
CATCHABILITY_6 * TOTAL_EFFORT_6 > CREEL6 THEN CREEL6 ELSE
CATCHABILITY_6 * TOTAL_EFFORT_6) ELSE IF ONOFF6 = 0 THEN
CATCHABILITY_6 * TOTAL_EFFORT_6 ELSE 0
GROWTH_6 = MALE150180 * GROWADULT
NAT_MORT6 = MALE150180 * NAT_MORT_RATE_6
MALE180210(t) = MALE180210(t − dt) + (GROWTH_6 − FISHING7 −
NATMORT7) * dt
INIT MALE180210 = 1000

INFLOWS:
GROWTH_6 = MALE150180 * GROWADULT
OUTFLOWS:
FISHING7 = IF ONOFF7 = 1 THEN (IF
CATCHABILITY_7 * TOTAL_EFFORT_7 > CREEL7 THEN CREEL7 ELSE
CATCHABILITY_7 * TOTAL_EFFORT_7) ELSE IF ONOFF7 = 0 THEN
CATCHABILITY_7 * TOTAL_EFFORT_7 ELSE 0
NATMORT7 = MALE180210 * NAT_MORT_RATE_7
MALE3060(t) = MALE3060(t − dt) + (GROWTH1 − NAT_MORT_2 −
GROWTH_2) * dt
INIT MALE3060 = 10000

INFLOWS:
GROWTH1 = GROW1 * MALE_030
OUTFLOWS:
NAT_MORT_2 = MALE3060 * NAT_MORT_RATE_2
GROWTH_2 = MALE3060 * GROW2
MALE6090(t) = MALE6090(t − dt) + (GROWTH_2 − NAT_MORT_3 −
GROWTH_3 − FISHING_3) * dt
INIT MALE6090 = 30000

INFLOWS:
GROWTH_2 = MALE3060 * GROW2
OUTFLOWS:
NAT_MORT_3 = NAT_MORT_RATE_3 * MALE6090
GROWTH_3 = MALE6090 * GROWADULT
FISHING_3 = IF ONOFF3 = 1 THEN (IF
CATCHABILITY_3 * TOTAL_EFFORT_3 > CREEL_3 THEN CREEL_3 ELSE
CATCHABILITY_3 * TOTAL_EFFORT_3) ELSE IF ONOFF3 = 0 THEN
CATCHABILITY_3 * TOTAL_EFFORT_3 ELSE 0
MALE90120(t) = MALE90120(t − dt) + (GROWTH_3 − FISHING_4 −
GROWTH_4 − NATMORT4) * dt

INIT MALE90120 = 50000

INFLOWS:
GROWTH_3 = MALE6090 * GROWADULT
OUTFLOWS:
FISHING_4 = IF ONOFF4 = 1 THEN (IF
CATCHABILITY_4 * TOTAL_EFFORT_4 > CREEL4 THEN CREEL4 ELSE
CATCHABILITY_4 * TOTAL_EFFORT_4) ELSE IF ONOFF4 = 0 THEN
CATCHABILITY_4 * TOTAL_EFFORT_4 ELSE 0
GROWTH_4 = MALE90120 * GROWADULT
NATMORT4 = MALE90120 * NAT_MORT_RATE4
MALE_030(t) = MALE_030(t − dt) + (PARENTAL_MALE −
NAT_MORT_1 − GROWTH1) * dt
INIT MALE_030 = 1000000

INFLOWS:
PARENTAL_MALE = MALE_FRY − SNEAKER
OUTFLOWS:
NAT_MORT_1 = NAT_MORT_RATE * MALE_030
GROWTH1 = GROW1 * MALE_030
MALE_FRY(t) = MALE_FRY(t − dt) + (MALE − PARENTAL_MALE −
SNEAKER) * dt
INIT MALE_FRY = 1000000

INFLOWS:
MALE = FRY * PROPORTIONMALE
OUTFLOWS:
PARENTAL_MALE = MALE_FRY − SNEAKER
SNEAKER = PROP_SN * MALE_FRY
SNEAKER_030(t) = SNEAKER_030(t − dt) + (SNEAKER −
SN_MORT_1 − SN_GROWTH_1) * dt
INIT SNEAKER_030 = 9000

INFLOWS:
SNEAKER = PROP_SN * MALE_FRY
OUTFLOWS:
SN_MORT_1 = SNEAKER_030 * SM_MORT_1_RATE
SN_GROWTH_1 = SN1GROW * SNEAKER_030
SN_3060(t) = SN_3060(t − dt) + (SN_GROWTH_1 − SN_GROWTH_2 −
SN_MORT_2) * dt
INIT SN_3060 = 1500

INFLOWS:
SN_GROWTH_1 = SN1GROW * SNEAKER_030
OUTFLOWS:
SN_GROWTH_2 = SN2GROW * SN_3060

SN_MORT_2 = SN_3060 * SM_MORT_2_RATE
SN_6090(t) = SN_6090(t − dt) + (SN_GROWTH_2 − SN_MORT_3) * dt
INIT SN_6090 = 500

INFLOWS:
SN_GROWTH_2 = SN2GROW * SN_3060
OUTFLOWS:
SN_MORT_3 = SN_6090 * SM_MORT_3_RATE
ADULT_DENSITY = ADULT_FEM_DENS + ADULT_MALE_DENS
ADULT_FEM_DENS = F_120150 + F_150180 + F_180210 + F_6090 +
F_90120
ADULT_MALE_DENS = MALE120150 + MALE150180 + MALE180210 +
MALE6090 + MALE90120
ANGLERS = 75
CREEL3 = 100
CREEL4 = 100
CREEL5 = 100
CREEL6 = 100
CREEL7 = 100
CREEL_3 = 100
CREEL_4 = 100
CREEL_5 = 100
CREEL_6 = 100
DISEASE = IF ADULT_DENSITY > 750000 * RANDOM(.8,1.2) THEN
RANDOM(1,3) ELSE 1
EGG_1 = F_90120 * 50000
EGG_2 = 50000 * F_120150
EGG_3 = F_150180 * 100000
EGG_4 = F_180210 * 100000
FRY_MORTALITY_RATE = .995 * DISEASE
F_CREEL_7 = 100
F_NATMORTRATE_3 = .0005 * DISEASE
F_NATMORTRATE_4 = .0005 * DISEASE
F_NATMORTRATE_5 = .0005 * DISEASE
F_NATMORTRATE_6 = .0005 * DISEASE
F_NATMORT_RATE_2 = .65 * DISEASE
F_NATMORT_RATE_7 = .9 * DISEASE
F_NATURAL_MORT_RATE_1 = .98 * DISEASE
F_ONOFF_3 = 1
F_ONOFF_4 = 1
F_ONOFF_5 = 1
F_ONOFF_6 = 1
F_ONOFF_7 = 1
F_TOT_EFFORT_3 = ANGLERS * ROD_HOURS
F_TOT_EFFORT_4 = ANGLERS * ROD_HOURS

F_TOT_EFFORT_5 = ANGLERS * ROD_HOURS
F_TOT_EFFORT_6 = ANGLERS * ROD_HOURS
F_TOT_EFFORT_7 = ANGLERS * ROD_HOURS
JUV_DENSITY = FEMALE_030 + F_3060 + MALE3060 + MALE_030
NAT_MORT_RATE = .98
NAT_MORT_RATE4 = .0005 * DISEASE
NAT_MORT_RATE_2 = .65 * DISEASE
NAT_MORT_RATE_3 = .0005 * DISEASE
NAT_MORT_RATE_5 = .0005 * DISEASE
NAT_MORT_RATE_6 = .0005 * DISEASE
NAT_MORT_RATE_7 = .9 * DISEASE
ONOFF3 = 1
ONOFF4 = 1
ONOFF5 = 1
ONOFF6 = 1
ONOFF7 = 1
OVERALL_EGG_PRODUCTION = EGG_4 + EGG_3 + EGG_2 + EGG_1
PROPORTIONMALE = 0.5
ROD_HOURS = 75
SM_MORT_1_RATE = 0.85 * DISEASE
SM_MORT_2_RATE = 0.10 * DISEASE
SM_MORT_3_RATE = 0.1 * DISEASE
TOTAL_EFFORT_3 = ANGLERS * ROD_HOURS
TOTAL_EFFORT_4 = ANGLERS * ROD_HOURS
TOTAL_EFFORT_5 = ANGLERS * ROD_HOURS
TOTAL_EFFORT_6 = ANGLERS * ROD_HOURS
TOTAL_EFFORT_7 = ANGLERS * ROD_HOURS
CATCHABILITY_3 = GRAPH(MALE6090)
(0.00, 0.004), (10000, 0.042), (20000, 0.062), (30000, 0.082), (40000, 0.128),
(50000, 0.158), (60000, 0.202), (70000, 0.258), (80000, 0.298), (90000, 0.342),
(100000, 0.386)
CATCHABILITY_4 = GRAPH(MALE90120)
(0.00, 0.048), (10000, 0.058), (20000, 0.078), (30000, 0.114), (40000, 0.15),
(50000, 0.194), (60000, 0.234), (70000, 0.274), (80000, 0.314), (90000, 0.346),
(100000, 0.394)
CATCHABILITY_5 = GRAPH(MALE120150)
(0.00, 0.086), (10000, 0.106), (20000, 0.134), (30000, 0.146), (40000, 0.174),
(50000, 0.226), (60000, 0.262), (70000, 0.29), (80000, 0.318), (90000, 0.34),
(100000, 0.4)
CATCHABILITY_6 = GRAPH(MALE150180)
(0.00, 0.162), (10000, 0.178), (20000, 0.206), (30000, 0.23), (40000, 0.262),
(50000, 0.29), (60000, 0.326), (70000, 0.342), (80000, 0.364), (90000, 0.376),
(100000, 0.4)
CATCHABILITY_7 = GRAPH(MALE180210)

(0.00, 0.356), (10000, 0.383), (20000, 0.477), (30000, 0.549), (40000, 0.585), (50000, 0.594), (60000, 0.626), (70000, 0.675), (80000, 0.761), (90000, 0.824), (100000, 0.9)

F_CATCH_3 = GRAPH(F_6090)

(0.00, 0.01), (10000, 0.036), (20000, 0.076), (30000, 0.108), (40000, 0.144), (50000, 0.176), (60000, 0.208), (70000, 0.244), (80000, 0.27), (90000, 0.326), (100000, 0.398)

F_CATCH_4 = GRAPH(F_90120)

(0.00, 0.03), (10000, 0.054), (20000, 0.082), (30000, 0.108), (40000, 0.144), (50000, 0.18), (60000, 0.224), (70000, 0.262), (80000, 0.322), (90000, 0.362), (100000, 0.4)

F_CATCH_5 = GRAPH(F_120150)

(0.00, 0.09), (10000, 0.128), (20000, 0.164), (30000, 0.18), (40000, 0.2), (50000, 0.212), (60000, 0.236), (70000, 0.272), (80000, 0.3), (90000, 0.332), (100000, 0.398)

F_CATCH_6 = GRAPH(F_150180)

(0.00, 0.174), (10000, 0.196), (20000, 0.22), (30000, 0.248), (40000, 0.27), (50000, 0.28), (60000, 0.302), (70000, 0.322), (80000, 0.336), (90000, 0.366), (100000, 0.396)

F_CATCH_7 = GRAPH(F_180210)

(0.00, 0.374), (10000, 0.414), (20000, 0.509), (30000, 0.594), (40000, 0.63), (50000, 0.662), (60000, 0.684), (70000, 0.725), (80000, 0.783), (90000, 0.815), (100000, 0.9)

GROW1 = GRAPH(JUV_DENSITY)

(0.00, 1.00), (400000, 0.91), (800000, 0.835), (1.2e + 06, 0.65), (1.6e + 06, 0.57), (2e + 06, 0.555), (2.4e + 06, 0.515), (2.8e + 06, 0.49), (3.2e + 06, 0.45), (3.6e + 06, 0.4), (4e + 06, 0.205)

GROW2 = GRAPH(ADULT_DENSITY)

(0.00, 0.98), (10000, 0.87), (20000, 0.77), (30000, 0.69), (40000, 0.67), (50000, 0.665), (60000, 0.655), (70000, 0.65), (80000, 0.59), (90000, 0.53), (100000, 0.475)

GROWADULT = GRAPH(ADULT_DENSITY)

(0.00, 1.00), (200000, 0.945), (400000, 0.915), (600000, 0.885), (800000, 0.865), (1e + 06, 0.845), (1.2e + 06, 0.82), (1.4e + 06, 0.72), (1.6e + 06, 0.67), (1.8e + 06, 0.625), (2e + 06, 0.59)

PROP_SN =

GRAPH(MALE120150 + MALE150180 + MALE180210 + MALE90120)

(0.00, 0.0187), (50000, 0.0248), (100000, 0.0315), (150000, 0.036), (200000, 0.0398), (250000, 0.045), (300000, 0.051), (350000, 0.0593), (400000, 0.0727), (450000, 0.0818), (500000, 0.0968)

SN1GROW = GRAPH(JUV_DENSITY)

(0.00, 0.445), (500000, 0.38), (1e + 06, 0.4), (1.5e + 06, 0.37), (2e + 06, 0.345), (2.5e + 06, 0.315), (3e + 06, 0.325), (3.5e + 06, 0.22), (4e + 06, 0.21), (4.5e + 06, 0.16), (5e + 06, 0.125)

SN2GROW = GRAPH(ADULT_DENSITY)

> (0.00, 0.4), (400000, 0.4), (800000, 0.375), (1.2e + 06, 0.345), (1.6e + 06, 0.33), (2e + 06, 0.315), (2.4e + 06, 0.285), (2.8e + 06, 0.26), (3.2e + 06, 0.235), (3.6e + 06, 0.21), (4e + 06, 0.18)

11.3 Wolly Adelgid

11.3.1 Infestation of Fraser Fir

Fraser fir populations have been decimated in the Mt. Mitchell area of Yancey County, North Carolina by the Balsam woolly adelgid, *Adelges piceae*. Fraser fir is a high altitude fir native to North America. *A. piceae* was introduced from Europe. Since its first detection in 1959, it has been responsible for the loss of over 2 million trees. Most impacted sites have been unable to recover fully. *A. piceae* is an external feeder on bark and new growth spirals. This feeding causes abnormal growth, eventually killing trees and adversely impacting the timber industry. A mature Fraser fir is estimated to support 50,000 A. piceae. Once this critical level is achieved, death of the tree is inevitable within two years. Death of a tree results in the loss of the pests on that tree from the entire park population.

In an attempt to prolong the life of the Fraser fir populations, 20 mature Fraser fir trees were moved from their natural habitat on Mt. Mitchell to a nearby city park on Grandfather Mountain, North Carolina. The city park will act as a botanic zoo producing a viable seed bank and maintaining genetic diversity in the Fraser fir. Unknowingly, a population of *Adelges piceae* was moved to the city park along with the Fraser fir. Unchecked, these insects may decimate the small population of Fraser fir moved to the park as part of a conservation measure.

Since the pest insect and host plant are now located in a city setting as opposed to a forest habitat, the options for control are much greater. However, the public perception of pesticide use in the city may make its use questionable politically. Considering the cost already invested to the Fraser fir conservation program, it is of great importance to preserve this city park population of Fraser fir. Using the model of *Adelges piceae* population dynamics and other associated model parameters, how does an uncontrolled population of *Adelges piceae* affect the *Fraser fir* population? When should pesticides be applied to ensure the greatest control over the pest population? How do varying levels of pesticide control affect the *Adelges piceae* and *Fraser fir* populations?

11.3.2 Adelgid and Fir Dynamics

The basis of this model is the population dynamics of the pest, *A. piceae,* modeled mathematically by[2]. The backbone of these dynamics is shown in Figures 11.17

---

[2] Dale, V.H., R. H. Gardner, and D.L. DeAngelis. 1991. Elevation-mediated effects of balsam woolly adelgid on southern Appalachian spruce-fir forests. Can. J. For. Res. 21: 1639–1648.

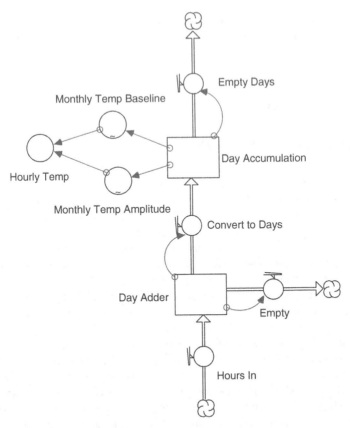

Fig. 11.17

through 11.19. Temperature is a large component to determining the development of the pest, so a temperature model (Figure 11.17) was built based on weather data from Grandfather Mountain, North Carolina[3].

Developmental rates of *A. piceae* have been previously modeled in the literature and are used here between life stages. Developmental rates were previously calculated in days and driven by temperature. Because of the habitat temperature fluctuations (altitude effect) this model includes daily temperature variation and developmental rates in hours. The temperature model thus gives hourly temperature on a 24-hour daily cycle. Changes in seasons are shown in Table 11.1.

Adelges piceae has five life stages—egg, adult, and three larval instars—and a stock are modeled for each stage (Figure 11.18). The adult reaches reproductive age in three weeks, so a conveyor is constructed for both pre- and reproductive adults. Development is modeled with flows from each stock, and egg laying is modeled with an inflow to the egg stock. *Adelges piceae* can overwinter only as first instar larvae,

---

[3] Washington Post. 1997. Weather Post, Grandfather Mountain, NC. http://www.weatherpost. com/ wp-srv/weath...orical/data/grandfather_mountain_nc.htm

Table 11.1

	Start date	End date
Winter	355	80
Spring	81	171
Summer	172	263
Fall	264	354

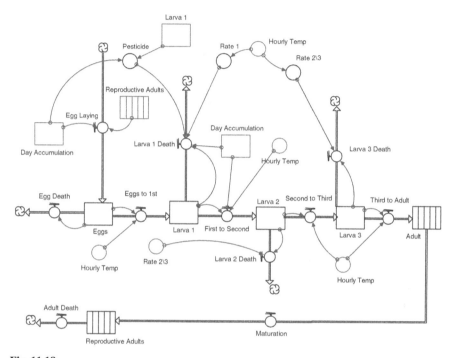

Fig. 11.18

and the model is currently evaluated based on an initial overwintering population of 10,000 first instar larvae. Death rates for each life stage have been constructed to simulate field population dynamics.

To model the effects of *Adelges piceae* on the Fraser fir population, we constructed a counter (Figure 11.19). The Sucker Adder counts the Total Suckers at each DT, and sums the change. Since each Fraser fir tree can host 50,000 *A. piceae* before it reaches death, the Sucker Adder was set to empty at this value. Thus, for every value of 50,000 *A. piceae*, a value of one was sent to the Sum of Damage stock, and in turn, one tree was infected.

The uninfected *Fraser fir* population began at 19 (1 infected tree arrived at our site), and this population quickly decreased as the uncontrolled *A. piceae* population grew and spread to new hosts (Figure 11.20).

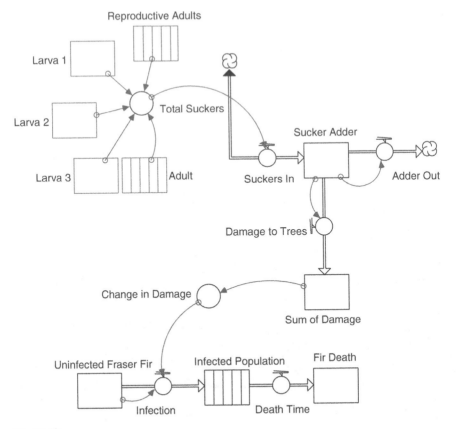

Fig. 11.19

As Figure 11.21 shows, the pest population grew exponentially. Without any pesticide control, *A. piceae* infected every tree within approximately 1.5 years. Since complete death of the tree takes between 1 and 2 years (accounted for by using the infected population conveyor), all of the trees would be decimated in less than 4 years. Given a longer running time, all 20 Fraser fir trees would be accounted for in the Fir Death stock.

Since biological control of the *Adelges piceae* has proven ineffective, chemical control is necessary to ensure a healthy Fraser fir population[4,5]. To model the effects of a pesticide control program, we altered the *Adelges piceae* death rate. By creating a pesticide "kill rate," we could easily manipulate the pest population. However,

[4] Mitchell, R.G. 1962. Balsam woolly aphid predators native to Oregon and Washington. Agricultural Experiment Station, Oregon State University, Technical Bulletin 62, 63pp.

[5] Hastings, F.L., F.P. Hain, A. Mangini, and W.T. Huxster. 1986. Control of the balsam woolly adelgid (Homoptera: Adelgidae) in Fraser fir Christmas tree plantations. J. Econ. Entomol. 79: 1676–1680.

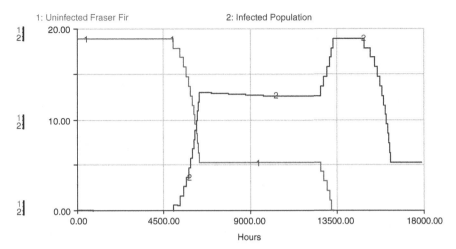

Fig. 11.20

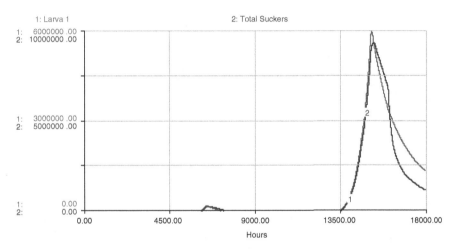

Fig. 11.21

before we could begin observing the affects of the pesticide, we had to determine the optimal application time. This was found to be on day 80, the last day of winter. In addition, the only remaining pests surviving the harsh conditions are those in the Larva 1 population. Thus, day 80 is the best time to apply pesticides because it is when the initial pest populations are the lowest. We experimented with different kill rates (which represent varying pesticide intensities) and found the minimum intensity of control necessary to maintain our original Fraser fir population. This is essential in a park setting, as this level will minimize environmental and other negative externalities. We experimented with the following pesticide kill rates: 98%, 11%, 10%, and 5% (Figures 11.22 through 11.25) and found that the optimal kill

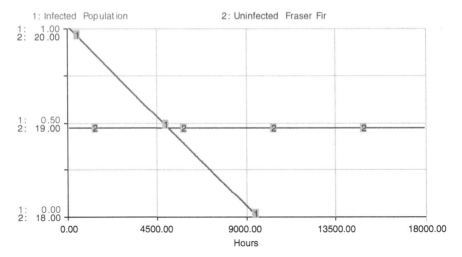

Fig. 11.22

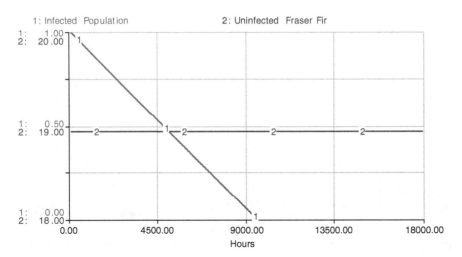

Fig. 11.23

rate for the desired results fell between 10% and 11%. At 11%, the Fraser fir population suffered no casualties (other than the originally infected tree). However, with a
10% kill rate, the Fraser fir population showed signs of steady decline and eventual
demise. This indicated that a 10% kill rate was not large enough to control our pest
population.

An 11% reduction appears to keep the *Adelges piceae* numbers relatively stable, as they are no longer infecting new hosts. Also, there is virtually no difference
between a 98% kill rate and a much lower one, such as 11%. In contrast, a 5%
kill rate would require an inadequate amount of pesticide, negatively affecting both
populations without significant gain on the pest control front.

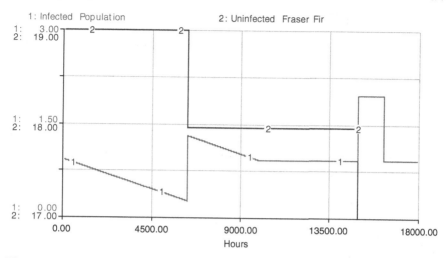

Fig. 11.24

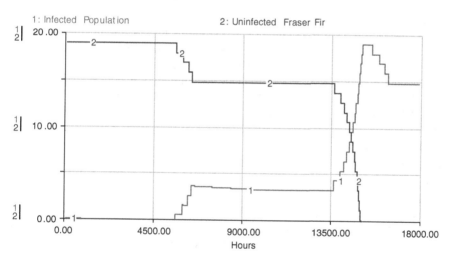

Fig. 11.25

11.3.3 Questions and Tasks

1. What are the environmental concerns of pesticide use (groundwater contamination, deleterious effects on other populations, and so forth) and how could those be included in the model?
2. What are the monetary costs of the modeled pesticide program? Are other methods of control more feasible (hand removal, temperature treatment, and other strategies)?

WOOLY ADELGID

Adult(t) = Adult(t − dt) + (Third_to_Adult − Maturation) * dt
  INIT Adult = 0
  TRANSIT TIME = 504
  INFLOW LIMIT = ∞
  CAPACITY = ∞

INFLOWS:
Third_to_Adult = Larva_3 * (1/(72 * (10.0 + 50 * (EXP(−0.25 *
(Hourly_Temp − 4))))))
OUTFLOWS:
Maturation = CONVEYOR OUTFLOW
Day_Accumulation(t) = Day_Accumulation(t − dt) + (Convert_to_Days −
Empty_Days) * dt
INIT Day_Accumulation = 0

INFLOWS:
Convert_to_Days = IF Day_Adder = 23 THEN 1 ELSE 0
OUTFLOWS:
Empty_Days = IF Day_Accumulation = 364 THEN Day_Accumulation ELSE IF
Day_Accumulation = 2 * 364 THEN Day_Accumulation ELSE IF
Day_Accumulation = 3 * 364 THEN Day_Accumulation ELSE IF
Day_Accumulation = 4 * 364 THEN Day_Accumulation ELSE IF
Day_Accumulation = 5 * 364 THEN Day_Accumulation ELSE 0
Day_Adder(t) = Day_Adder(t − dt) + (Hours_In − Convert_to_Days − Empty) *
dt
INIT Day_Adder = 0

INFLOWS:
Hours_In = 1
OUTFLOWS:
Convert_to_Days = IF Day_Adder = 23 THEN 1 ELSE 0
Empty = IF Day_Adder = 23 THEN Day_Adder ELSE 0
Eggs(t) = Eggs(t − dt) + (Egg_Laying − Eggs_to_1st − Egg_Death) * dt
INIT Eggs = 0

INFLOWS:
Egg_Laying = IF Day_Accumulation>=81 AND Day_Accumulation< = 171
THEN Reproductive_Adults * 10 ELSE IF Day_Accumulation> = 172 AND
Day_Accumulation<=263 THEN Reproductive_Adults * 5 ELSE 0
OUTFLOWS:
Eggs_to_1st = Eggs * (1/(24 * (5.0 + 100 * (EXP(−0.10 *
(Hourly_Temp − 4)))))))
Egg_Death = Eggs * 0.016
Fir_Death(t) = Fir_Death(t − dt) + (Death_Time) * dt
INIT Fir_Death = 0

INFLOWS:

Death_Time = CONVEYOR OUTFLOW

Infected_Population(t) = Infected_Population(t − dt) + (Infection − Death_Time) * dt

INIT Infected_Population = 1

TRANSIT TIME = 10000

INFLOW LIMIT = ∞

CAPACITY = ∞

INFLOWS:

Infection = IF (INT(Change_in_Damage > 0)) AND (Uninfected_Fraser_Fir > 0) THEN (1) ELSE 0

OUTFLOWS:

Death_Time = CONVEYOR OUTFLOW

Larva_1(t) = Larva_1(t − dt) + (Eggs_to_1st − First_to_Second − Larva_1_Death) * dt

INIT Larva_1 = 10000

INFLOWS:

Eggs_to_1st = Eggs * (1/(24 * (5.0 + 100 * (EXP(−0.10 * (Hourly_Temp − 4))))))

OUTFLOWS:

First_to_Second = IF Day_Accumulation>=226 AND Day_Accumulation<=80 THEN 0 ELSE Larva_1 * (1/(72 * (10.0 + 50 * (EXP(−0.25 * (Hourly_Temp − 4))))))

Larva_1_Death = IF (Day_Accumulation = 80) THEN Pesticide ELSE (Larva_1 * Rate_1)

Larva_2(t) = Larva_2(t − dt) + (First_to_Second − Second_to_Third − Larva_2_Death) * dt

INIT Larva_2 = 0

INFLOWS:

First_to_Second = IF Day_Accumulation>=226 AND Day_Accumulation<=80 THEN 0 ELSE Larva_1 * (1/(72 * (10.0 + 50 * (EXP(−0.25 * (Hourly_Temp − 4))))))

OUTFLOWS:

Second_to_Third = Larva_2 * (1/(72 * (10.0 + 50 * (EXP(−0.25 * (Hourly_Temp − 4))))))

Larva_2_Death = Larva_2 * Rate_2\3

Larva_3(t) = Larva_3(t − dt) + (Second_to_Third − Third_to_Adult − Larva_3_Death) * dt

INIT Larva_3 = 0

INFLOWS:

Second_to_Third = Larva_2 * (1/(72 * (10.0 + 50 * (EXP(−0.25 * (Hourly_Temp − 4))))))

OUTFLOWS:

Third_to_Adult = Larva_3 * (1/(72 * (10.0 + 50 * (EXP(−0.25 *
(Hourly_Temp − 4))))))
Larva_3_Death = Larva_3 * Rate_2\3
Reproductive_Adults(t) = Reproductive_Adults(t − dt) + (Maturation −
Adult_Death) * dt
INIT Reproductive_Adults = 0
   TRANSIT TIME = 240
   INFLOW LIMIT = ∞
   CAPACITY = ∞

INFLOWS:
Maturation = CONVEYOR OUTFLOW
OUTFLOWS:
Adult_Death = CONVEYOR OUTFLOW
Sucker_Adder(t) = Sucker_Adder(t − dt) + (Suckers_In − Adder_Out −
Damage_to_Trees) * dt
INIT Sucker_Adder = 10000

INFLOWS:
Suckers_In = (Total_Suckers − DELAY(Total_Suckers, DT))
OUTFLOWS:
Adder_Out = IF (Sucker_Adder >= 50000) THEN (50000 *
(INT(Sucker_Adder / 50000))) ELSE 0
Damage_to_Trees = IF (Sucker_Adder >= 50000) THEN (INT(Sucker_Adder /
50000)) ELSE 0
Sum_of_Damage(t) = Sum_of_Damage(t − dt) + (Damage_to_Trees) * dt
INIT Sum_of_Damage = 0

INFLOWS:
Damage_to_Trees = IF (Sucker_Adder >= 50000) THEN (INT(Sucker_Adder /
50000)) ELSE 0
Uninfected_Fraser_Fir(t) = Uninfected_Fraser_Fir(t − dt) + (−Infection) * dt
INIT Uninfected_Fraser_Fir = 19

OUTFLOWS:
Infection = IF (INT(Change_in_Damage > 0)) AND (Uninfected_Fraser_Fir > 0)
THEN (1) ELSE 0
Change_in_Damage = (Sum_of_Damage − DELAY(Sum_of_Damage, DT))
Hourly_Temp =
(Monthly_Temp_Amplitude * SIN(2 * PI * TIME/24)) +
Monthly_Temp_Baseline
Pesticide = IF (Day_Accumulation = 80) THEN (Larva_1 * .11) ELSE 0
Rate_1 = IF Hourly_Temp<=4 THEN 0.0003 ELSE 0.0001
Rate_2\3 = IF Hourly_Temp<=4 THEN 1/72 ELSE 0.0001
Total_Suckers = Adult + Larva_1 + Larva_2 + Larva_3 + Reproductive_Adults
Monthly_Temp_Amplitude = GRAPH(Day_Accumulation)

(0.00, 4.50), (1.00, 4.50), (2.01, 4.50), (3.01, 4.50), (4.01, 4.50), (5.01, 4.50), (6.02, 4.50), (7.02, 4.50), (8.02, 4.50), (9.02, 4.50), (10.0, 4.50), (11.0, 4.50), (12.0, 4.50), (13.0, 4.50), (14.0, 4.50), (15.0, 4.50), (16.0, 4.50), (17.0, 4.50), (18.0, 4.50), (19.1, 4.50), (20.1, 4.50), (21.1, 4.50), (22.1, 4.50), (23.1, 4.50), (24.1, 4.50), (25.1, 4.50), (26.1, 4.50), (27.1, 4.50), (28.1, 4.50), (29.1, 4.50), (30.1, 4.50), (31.1, 4.50), (32.1, 4.50), (33.1, 4.50), (34.1, 4.50), (35.1, 4.50), (36.1, 4.50), (37.1, 4.50), (38.1, 4.50), (39.1, 4.50), (40.1, 4.50), (41.1, 4.50), (42.1, 4.50), (43.1, 4.50), (44.1, 4.50), (45.1, 4.50), (46.1, 4.50), (47.1, 4.50), (48.1, 4.50), (49.1, 4.50), (50.1, 4.50), (51.1, 4.50), (52.1, 4.50), (53.1, 4.50), (54.1, 4.50), (55.2, 4.50), (56.2, 4.50), (57.2, 4.50), (58.2, 4.50), (59.2, 4.50), (60.2, 4.50), (61.2, 4.50), (62.2, 4.50), (63.2, 4.50), (64.2, 4.50), (65.2, 4.50), (66.2, 4.50), (67.2, 4.50), (68.2, 4.50), (69.2, 4.50), (70.2, 4.50), (71.2, 4.50), (72.2, 4.50), (73.2, 4.50), (74.2, 4.50), (75.2, 4.50), (76.2, 4.50), (77.2, 4.50), (78.2, 4.50), (79.2, 4.50), (80.2, 4.50), (81.2, 4.50), (82.2, 4.50), (83.2, 4.50), (84.2, 4.50), (85.2, 4.50), (86.2, 4.50), (87.2, 4.50), (88.2, 4.50), (89.2, 4.50), (90.2, 4.50), (91.3, 5.00), (92.3, 5.00), (93.3, 5.00), (94.3, 5.00), (95.3, 5.00), (96.3, 5.00), (97.3, 5.00), (98.3, 5.00), (99.3, 5.00), (100, 5.00), (101, 5.00), (102, 5.00), (103, 5.00), (104, 5.00), (105, 5.00), (106, 5.00), (107, 5.00), (108, 5.00), (109, 5.00), (110, 5.00), (111, 5.00), (112, 5.00), (113, 5.00), (114, 5.00), (115, 5.00), (116, 5.00), (117, 5.00), (118, 5.00), (119, 5.00), (120, 5.00), (121, 4.50), (122, 4.50), (123, 4.50), (124, 4.50), (125, 4.50), (126, 4.50), (127, 4.50), (128, 4.50), (129, 4.50), (130, 4.50), (131, 4.50), (132, 4.50), (133, 4.50), (134, 4.50), (135, 4.50), (136, 4.50), (137, 4.50), (138, 4.50), (139, 4.50), (140, 4.50), (141, 4.50), (142, 4.50), (143, 4.50), (144, 4.50), (145, 4.50), (146, 4.50), (147, 4.50), (148, 4.50), (149, 4.50), (150, 4.50), (151, 4.50), (152, 4.50), (153, 4.00), (154, 4.00), (155, 4.00), (156, 4.00), (157, 4.00), (158, 4.00), (159, 4.00), (160, 4.00), (161, 4.00), (162, 4.00), (163, 4.00), (164, 4.00), (165, 4.00), (166, 4.00), (167, 4.00), (168, 4.00), (169, 4.00), (170, 4.00), (171, 4.00), (172, 3.50), (173, 3.50), (174, 3.50), (175, 3.50), (176, 3.50), (177, 3.50), (178, 3.50), (179, 3.50), (180, 3.50), (181, 3.50), (183, 3.50), (184, 3.50), (185, 3.50), (186, 3.50), (187, 3.50), (188, 3.50), (189, 3.50), (190, 3.50), (191, 3.50), (192, 3.50), (193, 3.50), (194, 3.50), (195, 3.50), (196, 3.50), (197, 3.50), (198, 3.50), (199, 3.50), (200, 3.50), (201, 3.50), (202, 3.50), (203, 3.50), (204, 3.50), (205, 3.50), (206, 3.50), (207, 3.50), (208, 3.50), (209, 3.50), (210, 3.50), (211, 3.50), (212, 3.50), (213, 3.50), (214, 3.50), (215, 3.50), (216, 3.50), (217, 3.50), (218, 3.50), (219, 3.50), (220, 3.50), (221, 3.50), (222, 3.50), (223, 3.50), (224, 3.50), (225, 3.50), (226, 3.50), (227, 3.50), (228, 3.50), (229, 3.50), (230, 3.50), (231, 3.50), (232, 3.50), (233, 3.50), (234, 3.50), (235, 3.50), (236, 3.50), (237, 3.50), (238, 3.50), (239, 3.50), (240, 3.50), (241, 3.50), (242, 3.50), (243, 3.50), (244, 3.50), (245, 3.50), (246, 3.50), (247, 3.50), (248, 3.50), (249, 3.50), (250, 3.50), (251, 3.50), (252, 3.50), (253, 3.50), (254, 3.50), (255, 3.50), (256, 3.50), (257, 3.50), (258, 3.50), (259, 3.50), (260, 3.50), (261, 3.50), (262, 3.50), (263, 3.50), (264, 3.50), (265, 3.50), (266, 3.50), (267, 3.50), (268, 4.00), (269, 4.00), (270, 4.00), (271, 4.00), (272, 4.00), (273, 4.00), (274, 4.00), (275, 4.00), (276, 4.00), (277, 4.00), (278, 4.00), (279, 4.00), (280, 4.00), (281, 4.00), (282, 4.00), (283, 4.00), (284, 4.00), (285,

4.00), (286, 4.00), (287, 4.00), (288, 4.00), (289, 4.00), (290, 4.00), (291, 4.00), (292, 4.00), (293, 4.00), (294, 4.00), (295, 4.00), (296, 4.00), (297, 4.00), (298, 4.00), (299, 4.00), (300, 4.00), (301, 4.00), (302, 4.00), (303, 4.00), (304, 4.00), (305, 4.00), (306, 4.00), (307, 4.00), (308, 4.00), (309, 4.00), (310, 4.00), (311, 4.00), (312, 4.00), (313, 4.00), (314, 4.00), (315, 4.00), (316, 4.00), (317, 4.00), (318, 4.00), (319, 4.00), (320, 4.00), (321, 4.00), (322, 4.00), (323, 4.00), (324, 4.00), (325, 4.00), (326, 4.00), (327, 4.00), (328, 4.00), (329, 3.00), (330, 3.00), (331, 3.00), (332, 3.00), (333, 3.00), (334, 3.00), (335, 3.00), (336, 3.00), (337, 3.00), (338, 3.00), (339, 3.00), (340, 3.00), (341, 3.00), (342, 3.00), (343, 3.00), (344, 3.00), (345, 3.00), (346, 3.00), (347, 3.00), (348, 3.00), (349, 3.00), (350, 3.00), (351, 3.00), (352, 3.00), (353, 3.00), (354, 3.00), (355, 3.00), (356, 3.00), (357, 3.00), (358, 3.00), (359, 3.00), (360, 3.00), (361, 3.00), (362, 3.00), (363, 3.00), (364, 3.00)

Monthly_Temp_Baseline = GRAPH(Day_Accumulation)

(0.00, −2.50), (1.00, −2.50), (2.01, −2.50), (3.01, −2.50), (4.01, −2.50), (5.01, −2.50), (6.02, −2.50), (7.02, −2.50), (8.02, −2.50), (9.02, −2.50), (10.0, −2.50), (11.0, −2.50), (12.0, −2.50), (13.0, −2.50), (14.0, −2.50), (15.0, −2.50), (16.0, −2.50), (17.0, −2.50), (18.0, −2.50), (19.1, −2.50), (20.1, −2.50), (21.1, −2.50), (22.1, −2.50), (23.1, −2.50), (24.1, −2.50), (25.1, −2.50), (26.1, −2.50), (27.1, −2.50), (28.1, −2.50), (29.1, −2.50), (30.1, −2.50), (31.1, −2.50), (32.1, −1.50), (33.1, −1.50), (34.1, −1.50), (35.1, −1.50), (36.1, −1.50), (37.1, −1.50), (38.1, −1.50), (39.1, −1.50), (40.1, −1.50), (41.1, −1.50), (42.1, −1.50), (43.1, −1.50), (44.1, −1.50), (45.1, −1.50), (46.1, −1.50), (47.1, −1.50), (48.1, −1.50), (49.1, −1.50), (50.1, −1.50), (51.1, −1.50), (52.1, −1.50), (53.1, −1.50), (54.1, −1.50), (55.2, −1.50), (56.2, −1.50), (57.2, −1.50), (58.2, −1.50), (59.2, −1.50), (60.2, 2.50), (61.2, 2.50), (62.2, 2.50), (63.2, 2.50), (64.2, 2.50), (65.2, 2.50), (66.2, 2.50), (67.2, 2.50), (68.2, 2.50), (69.2, 2.50), (70.2, 2.50), (71.2, 2.50), (72.2, 2.50), (73.2, 2.50), (74.2, 2.50), (75.2, 2.50), (76.2, 2.50), (77.2, 2.50), (78.2, 2.50), (79.2, 2.50), (80.2, 2.50), (81.2, 2.50), (82.2, 2.50), (83.2, 2.50), (84.2, 2.50), (85.2, 2.50), (86.2, 2.50), (87.2, 2.50), (88.2, 2.50), (89.2, 2.50), (90.2, 2.50), (91.3, 7.00), (92.3, 7.00), (93.3, 7.00), (94.3, 7.00), (95.3, 7.00), (96.3, 7.00), (97.3, 7.00), (98.3, 7.00), (99.3, 7.00), (100, 7.00), (101, 7.00), (102, 7.00), (103, 7.00), (104, 7.00), (105, 7.00), (106, 7.00), (107, 7.00), (108, 7.00), (109, 7.00), (110, 7.00), (111, 7.00), (112, 7.00), (113, 7.00), (114, 7.00), (115, 7.00), (116, 7.00), (117, 7.00), (118, 7.00), (119, 7.00), (120, 7.00), (121, 11.5), (122, 11.5), (123, 11.5), (124, 11.5), (125, 11.5), (126, 11.5), (127, 11.5), (128, 11.5), (129, 11.5), (130, 11.5), (131, 11.5), (132, 11.5), (133, 11.5), (134, 11.5), (135, 11.5), (136, 11.5), (137, 11.5), (138, 11.5), (139, 11.5), (140, 11.5), (141, 11.5), (142, 11.5), (143, 11.5), (144, 11.5), (145, 11.5), (146, 11.5), (147, 11.5), (148, 11.5), (149, 11.5), (150, 11.5), (151, 11.5), (152, 15.0), (153, 15.0), (154, 15.0), (155, 15.0), (156, 15.0), (157, 15.0), (158, 15.0), (159, 15.0), (160, 15.0), (161, 15.0), (162, 15.0), (163, 15.0), (164, 15.0), (165, 15.0), (166, 15.0), (167, 15.0), (168, 15.0), (169, 15.0), (170, 15.0), (171, 15.0), (172, 15.0), (173, 15.0), (174, 15.0), (175, 15.0), (176, 15.0), (177, 15.0), (178, 15.0), (179, 15.0), (180, 15.0), (181,

16.5), (183, 16.5), (184, 16.5), (185, 16.5), (186, 16.5), (187, 16.5), (188, 16.5), (189, 16.5), (190, 16.5), (191, 16.5), (192, 16.5), (193, 16.5), (194, 16.5), (195, 16.5), (196, 16.5), (197, 16.5), (198, 16.5), (199, 16.5), (200, 16.5), (201, 16.5), (202, 16.5), (203, 16.5), (204, 16.5), (205, 16.5), (206, 16.5), (207, 16.5), (208, 16.5), (209, 16.5), (210, 16.5), (211, 16.5), (212, 16.5), (213, 16.5), (214, 16.5), (215, 16.5), (216, 16.5), (217, 16.5), (218, 16.5), (219, 16.5), (220, 16.5), (221, 16.5), (222, 16.5), (223, 16.5), (224, 16.5), (225, 16.5), (226, 16.5), (227, 16.5), (228, 16.5), (229, 16.5), (230, 16.5), (231, 16.5), (232, 16.5), (233, 16.5), (234, 16.5), (235, 16.5), (236, 16.5), (237, 16.5), (238, 16.5), (239, 16.5), (240, 16.5), (241, 16.5), (242, 16.5), (243, 13.5), (244, 13.5), (245, 13.5), (246, 13.5), (247, 13.5), (248, 13.5), (249, 13.5), (250, 13.5), (251, 13.5), (252, 13.5), (253, 13.5), (254, 13.5), (255, 13.5), (256, 13.5), (257, 13.5), (258, 13.5), (259, 13.5), (260, 13.5), (261, 13.5), (262, 13.5), (263, 13.5), (264, 13.5), (265, 13.5), (266, 13.5), (267, 13.5), (268, 13.5), (269, 13.5), (270, 13.5), (271, 13.5), (272, 8.00), (273, 8.00), (274, 8.00), (275, 8.00), (276, 8.00), (277, 8.00), (278, 8.00), (279, 8.00), (280, 8.00), (281, 8.00), (282, 8.00), (283, 8.00), (284, 8.00), (285, 8.00), (286, 8.00), (287, 8.00), (288, 8.00), (289, 8.00), (290, 8.00), (291, 8.00), (292, 8.00), (293, 8.00), (294, 8.00), (295, 8.00), (296, 8.00), (297, 8.00), (298, 8.00), (299, 8.00), (300, 8.00), (301, 8.00), (302, 8.00), (303, 4.00), (304, 4.00), (305, 4.00), (306, 4.00), (307, 4.00), (308, 4.00), (309, 4.00), (310, 4.00), (311, 4.00), (312, 4.00), (313, 4.00), (314, 4.00), (315, 4.00), (316, 4.00), (317, 4.00), (318, 4.00), (319, 4.00), (320, 4.00), (321, 4.00), (322, 4.00), (323, 4.00), (324, 4.00), (325, 4.00), (326, 4.00), (327, 4.00), (328, 4.00), (329, 4.00), (330, 4.00), (331, 4.00), (332, 4.00), (333, 0.00), (334, 0.00), (335, 0.00), (336, 0.00), (337, 0.00), (338, 0.00), (339, 0.00), (340, 0.00), (341, 0.00), (342, 0.00), (343, 0.00), (344, 0.00), (345, 0.00), (346, 0.00), (347, 0.00), (348, 0.00), (349, 0.00), (350, 0.00), (351, 0.00), (352, 0.00), (353, 0.00), (354, 0.00), (355, 0.00), (356, 0.00), (357, 0.00), (358, 0.00), (359, 0.00), (360, 0.00), (361, 0.00), (362, 0.00), (363, 0.00), (364, 0.00)

Chapter 12
Indirect Susceptible-Infected-Resistant Models of Arboviral Encephalitis Transmission[*]

12.1 Modeling West Nile Virus Dynamics Emily Wheeler and Traci Barkley

West Nile virus (WNV) is an arthropod borne, or arboviral, disease historically endemic to Africa. In Africa, the disease circulates between populations of wild birds and mosquitoes, but only occasionally results in significant disease outbreaks in wildlife[1]. After emerging in North America in 1999, however, WNV spread rapidly through avian communities, causing unexpected mortality in many bird species and encephalitis epidemics in secondary hosts, such as horses and humans. WNV is one of a number of related flaviviral encephalitides, including eastern equine encephalitis, Japanese encephalitis, and La Cross encephalitis. These diseases are all transmitted among hosts by mosquitoes and can result in symptoms ranging from mild malaise to severe neurological disease and death[2].

Dynamic modeling of emerging diseases like WNV benefits understanding of the diverse interactions of EXTRINSIC drivers, such as weather, seasonal demographics or habitat types, on INTRINSIC host and vector interactions that determine the severity, location, and timing of disease amplification and transmission (Figure 12.1). While complicated by these extrinsic drivers, the intrinsic components of an arboviral encephalitis system like WNV can be reasonably described by extension of a classical epidemiological model, the common Susceptible-Infectious-Resistant (SIR) model. With a few simple changes, this model can be extended to accommodate indirect disease transmission among two or more interacting populations of hosts and vectors. After building this base indirect SIR model, extrinsic drivers can be added to show how the outbreak might change with this increased complexity.

---

[*] This model was developed by Emily Wheeler and Tracy Barkley.

[1] Jupp, P. 2001. The ecology of West Nile virus in South Africa and the occurrence of outbreaks in humans. West Nile Virus: Detection, Surveillance, and Control Annals of the New York Academy of Science. 951: 143–152.

[2] Center for Disease Control and Prevention. West Nile Virus [cited 2007 September 9]. Available from http://www.cdc.gov/ncidod/dvbid/westnile/index.htm.

B. Hannon and M. Ruth, *Dynamic Modeling of Diseases and Pests*,
Modeling Dynamic Systems,
© Springer Science + Business Media LLC 2009

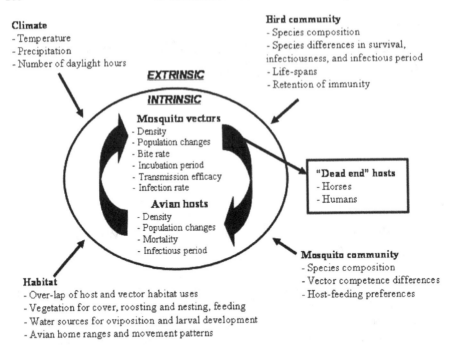

Climate
- Temperature
- Precipitation
- Number of daylight hours

Bird community
- Species composition
- Species differences in survival, infectiousness, and infectious period
- Life-spans
- Retention of immunity

EXTRINSIC

INTRINSIC

Mosquito vectors
- Density
- Population changes
- Bite rate
- Incubation period
- Transmission efficacy
- Infection rate

Avian hosts
- Density
- Population changes
- Mortality
- Infectious period

"Dead end" hosts
- Horses
- Humans

Mosquito community
- Species composition
- Vector competence differences
- Host-feeding preferences

Habitat
- Over-lap of host and vector habitat uses
- Vegetation for cover, roosting and nesting, feeding
- Water sources for oviposition and larval development
- Avian home ranges and movement patterns

Fig. 12.1

In this chapter we build from a simple, population-dynamic SIR model of a contagious disease to an indirectly transmitted arboviral disease model with explicit population and seasonal dynamics. We use these models to qualitatively explore how host and vector natural history and population dynamics might change predictions as to how the disease outbreak will behave over time.

12.2 Susceptible-Infected-Resistant (SIR) Models in Dynamic Populations

12.2.1 Model Structure and Behavior

[3]established the value of including host population dynamics in models of bacterial and viral disease outbreaks. They first incorporated changes in population density to the classical SIR model, allowing for more ecologically based modeling of disease outbreaks in changing populations. While long used in a population-static form

---

[3] Anderson, R.M. and R.M. May. Population biology of infectious diseases: Part I. Nature 1979a; 280: 361–367.

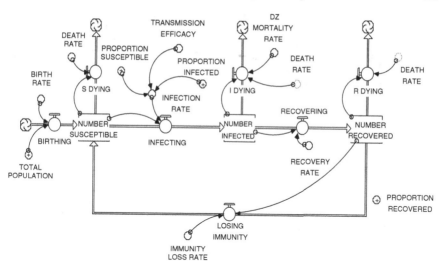

Fig. 12.2

for the study of limited disease outbreaks, expansion of the SIR model to dynamically changing populations allowed for the study of more complex disease dynamics where addition of new susceptible individuals can change predictions about disease persistence (Figure 12.2).

Whether population-dynamic or population-static, in an SIR model each stock holds the number of individuals of a certain disease status—the NUMBER SUSCEPTIBLE to infection with the disease, the NUMBER INFECTED and capable of passing the disease to others, and the NUMBER RECOVERED from the disease (a state often assumed to be immune to re-infection for some period of time or permanently, depending on the disease). Rates of transition between susceptible and infected status depend on the TRANSMISSION EFFICACY—or what proportion of contacts between susceptible and infected individuals result in transmission of the disease—and the rate of contact between infected and susceptible individuals, which is determined by multiplying the proportions of individuals in the INFECTED and SUSCEPTIBLE categories. BIRTHING and DYING are built into the model as a simple density independent growth equation where each compartment has a background DEATH RATE. In this model, we assume a disease that causes only morbidity (not mortality) so the DZ MORTALITY RATE is set to zero. Should we wish to model a fatal disease, then the death rate for the infected individuals would be elevated from the population's average DEATH RATE to the background DEATH RATE plus the DZ MORTALITY RATE.

Some disease outbreaks may be temporally or spatially limited such that population level changes have little effect on the epidemic. For example, an influenza outbreak at a convalescent center or hospital may reasonably be modeled in terms of a closed population of potentially exposed individuals who transition through susceptible, infected, and resistant status. The outbreak ends when the relative proportion

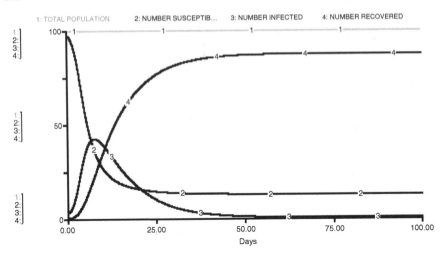

Fig. 12.3

of susceptible individuals is low enough that contacts between infected individuals and the remaining susceptible individuals are rare enough that transmission declines to zero. If we run our simple dynamic SIR model with both the BIRTH RATE and DEATH RATE set to zero, we get the classical population-static form of the model, which produces a classical epidemiological curve for a limited disease outbreak (Figure 12.3). Note how the TOTAL POPULATION (light gray) stays constant and the outbreak ends when the majority of the population is immune to the disease (i.e. majority of individuals are RECOVERED).

However, many infectious diseases occur over time periods in which the addition of new susceptible individuals through births might drastically alter the course of the disease. What happens to the epidemic curve when the population is no longer static, but growing steadily? (Figure 12.4). With the constant addition of new susceptible individuals, the disease persists in the population instead of going "extinct," as in the previous run of the model. Note how population size (light grey) is increasing with a slow exponential curve. In the population-static model the disease disappears around day 50. In the population-dynamic SIR model, however, a disease can cycle or become a persistent element where the model reaches a stable equilibrium in which the disease persists at low prevalence, depending on the rate of population growth.

12.2.2 Questions and Tasks

As we saw in this section, population dynamics can alter the course of the disease, resulting in persistence in the population.

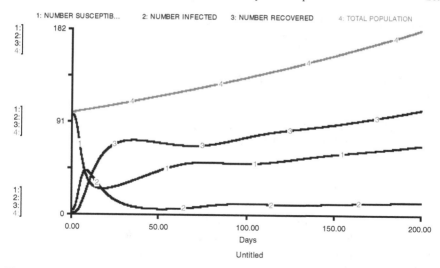

Untitled

Fig. 12.4

1. What other factor(s) might also contribute to persistence of a disease in a population? (Hint: How might you model an evolving pathogen that can rapidly re-infect and cause disease in previously exposed individuals?)
2. How would you model migration of new susceptible individuals from a neighboring population and what impact would migration have on the outbreak dynamics?

SIMPLE POPULATION-DYNAMIC SIR MODEL

NUMBER_INFECTED(t) = NUMBER_INFECTED(t − dt) + (INFECTING − RECOVERING − I_DYING) * dt
INIT NUMBER_INFECTED = 3

INFLOWS:
INFECTING = NUMBER_SUSCEPTIBLE * INFECTION_RATE
OUTFLOWS:
RECOVERING = NUMBER_INFECTED * RECOVERY_RATE
I_DYING = DEATH_RATE*NUMBER_INFECTED + DZ_MORTALITY_RATE*NUMBER_INFECTED
NUMBER_RECOVERED(t) = NUMBER_RECOVERED(t − dt) + (RECOVERING − LOSING_IMMUNITY − R_DYING) * dt
INIT NUMBER_RECOVERED = 0

INFLOWS:
RECOVERING = NUMBER_INFECTED * RECOVERY_RATE
OUTFLOWS:
LOSING_IMMUNITY = NUMBER_RECOVERED*IMMUNITY_LOSS_RATE
R_DYING = NUMBER_RECOVERED * DEATH_RATE
NUMBER_SUSCEPTIBLE(t) = NUMBER_SUSCEPTIBLE(t − dt) +

(LOSING_IMMUNITY + BIRTHING − INFECTING − S_DYING) * dt
INIT NUMBER_SUSCEPTIBLE = 97

INFLOWS:
LOSING_IMMUNITY = NUMBER_RECOVERED*IMMUNITY_LOSS_RATE
BIRTHING = BIRTH_RATE * TOTAL_POPULATION
OUTFLOWS:
INFECTING = NUMBER_SUSCEPTIBLE * INFECTION_RATE
S_DYING = NUMBER_SUSCEPTIBLE * DEATH_RATE
BIRTH_RATE = 0.014
DEATH_RATE = 0.011
DZ_MORTALITY_RATE = 0
IMMUNITY_LOSS_RATE = 0
INFECTION_RATE = PROPORTION_INFECTED *
PROPORTION_SUSCEPTIBLE * TRANSMISSION_EFFICACY
PROPORTION_INFECTED = NUMBER_INFECTED/TOTAL_POPULATION
PROPORTION_RECOVERED =
NUMBER_RECOVERED/TOTAL_POPULATION
PROPORTION_SUSCEPTIBLE =
NUMBER_SUSCEPTIBLE/TOTAL_POPULATION
RECOVERY_RATE = 1/8
TOTAL_POPULATION = NUMBER_SUSCEPTIBLE +
NUMBER_RECOVERED + NUMBER_INFECTED
TRANSMISSION_EFFICACY = 1

12.3 Base WNV SIR Model with a Dynamic Vector Population

12.3.1 Base Model Structure and Behavior

With a few simple modifications, a population-dynamic SIR model can also be applied to diseases with indirect transmission through a vector[4]. Complexity certainly increases when we consider dynamically changing host and vector population densities, even as these populations interact with one another through disease transmission feedbacks. Further, the disease transmission cycle may negatively feedback on population growth rates, reducing reproductive output of infected or even recovered individuals. This is especially true for the host population if illness can result in permanent or slow-healing damages, such as chronic neurological deficits, which may alter reproductive behaviors or physiology. In this chapter, we will make the simplifying, but unrealistic assumption that neither the host nor the vector reproductive

---

[4] Anderson, R.M. and R.M. May. Population biology of infectious diseases: Part II. Nature 1979b; 280: 455–461.

rates are altered by disease status, but this type of feedback certainly has interesting implications to consider (e.g. see [5]).

For a contagious disease, transmission rate depends on the probability of a susceptible individual coming into close contact with an infected individual and the efficacy of those interactions in relaying the disease. However, vector transmitted diseases like WNV have at least two separate populations that transition through the stages of disease exposure via a positive feedback loop between susceptible individuals of each population with the infected individuals of the other population. Overlying this feedback loop are the intrinsic population dynamics that contribute new susceptible individuals to the host and vector populations. Adding a final layer of complexity is the fact that hosts and vectors often have highly divergent reproductive time scales, with vector populations turning over many times a month while their avian targets may suffer the bites of these generations of vectors over the course of months to years. How might these different time scales affect disease transmission dynamics over short and long time scales?

[6]published a model that describes a model of the intrinsic factors in WNV, focusing on the interactions of a static avian population and a dynamic vector population within a single season. This eight-compartment model includes avian hosts transitioning among BIRD SUSCEPTIBLE, BIRD INFECTED, BIRD RECOVERED, and BIRD DEAD with a static population structure that ignores births and deaths outside of disease mortality. The mosquito vector population is modeled with a combined demographic and disease transition structure with production of new LARVAL individuals by existing adults of all categories. They transition via development to the adult MOSQUITO SUSCEPTIBLE. It is assumed that there is no vertical transmission of the disease from an infected female mosquito to larva through the ovary in this model. Adult mosquitoes transition from MOSQUITO SUSCEPTIBLE to MOSQUITO EXPOSED and then to MOSQUITO INFECTED status, at which time they can transmit the disease back to BIRD SUSCEPTIBLE individuals. The MOSQUITO EXPOSED stock represents mosquitoes that have bitten an infected host, but are not yet competent to transmit the disease. Transition from MOSQUITO SUSCEPTIBLE to MOSQUITO EXPOSED relies on encountering and biting an infected individual, while transition from MOSQUITO EXPOSED to MOSQUITO INFECTED relies on amplification of the virus within the vector to levels at which the disease is transmitted during feeding on the host.

Parameter values for this model general follow from the model based on American crows developed by.[6] However, as the goal of this chapter is not to quantitatively analyze WNV *per se* but rather to explore the interactions that may

[5] Dobson, A., I. Cattadori, R.D. Holt, R.S. Ostfeld, F. Keesing, K. Krichbaum, J.R. Rohr, S.E. Perkins, and P.J. Hudson. 2006. Sacred cows and sympathetic squirrels: The importance of biological diversity to human health. PLOS Medicine. 3: 714–718.

[6] Wonham, M., T. de-Camino-Beck, and M. Lewis. 2004. An Epidemiological model for West Nile Virus: invasion analysis and control applications. Proceedings of the Royal Society of London. 271: 501–507.

drive observed patterns in this disease system, some parameters were varied within the ranges provided by[6] to help clarify interesting patterns that emerge from these models.

The RECOVERY RATE, for example, was set to 0.05 individuals recovering per day to generalize the model to the broader avian community. Unlike the American crow, in which the disease has a high fatality rate[7], other avian species demonstrate variable survival[8]. The BITE RATE was set to the high end of the estimated range for American crows,[6] which is still likely an underestimation of the total bite rate on a complete avian community. Finally, unlike in the original model, a seasonal "on-off" switch (using an if–then statement) was added to the BITE RATE, MOSQUITO DEATH RATE, and OVIPOSITION RATE to qualitatively mimic the basic seasonal dynamics during which temperature alters annual mosquito activity. Figures 12.5 and 12.6 demonstrate the structure of this base model in STELLA for the host and vector, respectively.

WNV has been shown to persist in overwintering mosquitoes, and many researchers believe that this contributes to the persistence of the virus through the

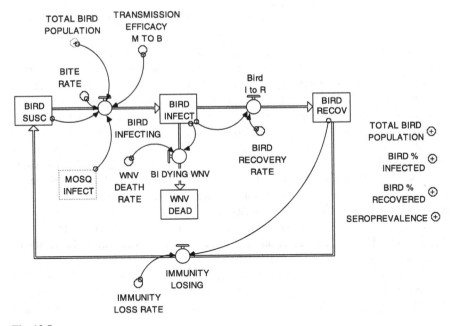

Fig. 12.5

[7] Yaremych, S., R. Warner, P. Mankin, J. Brawn, A. Raim, and R. Novak. 2004. West Nile Virus and High Death Rate in American Crows. Emerging Infectious Diseases. 10(4): 709–711.

[8] Komar, N., Langevin, S. Hinten, S., Nemeth, N. Edwards, E., and D. Hettler et al. 2003. Experimental Infection of North American Birds with the New York 1999 strain of West Nile Virus. Emerging Infectious Diseases. 9: 311–22.

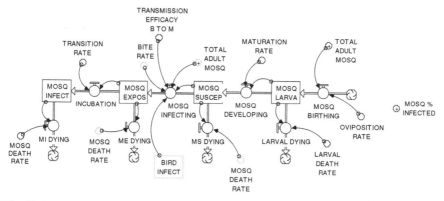

Fig. 12.6

winters in northern areas of the United States[9]. Using estimates of the proportion of overwintering mosquitoes infected with WNV to initiate the transmission cycle, the small PERCENT MOSQUITOES INFECTED in the overwintering mosquito population (which were used to initiate the transmission cycle) begins to rise in early summer when the mosquitoes emerge to begin breeding and biting hosts. This rise is accompanied by a rise in the PERCENT INFECTED BIRDS, or disease prevalence in the host population (Figure 12.7). Since very few studies of WNV have the resources to test birds directly for viremia, we can also evaluate the simulated outbreak in terms of SEROPREVALENCE, or the proportion of individuals that test positive for disease exposure by the presence of antibodies to the disease in their bloodstream. This value includes not only currently infected individuals, but also recovered individuals who can no longer transmit the disease but have been exposed at some time in the past.

12.3.2 Questions and Tasks

It has been proposed that vertical trans-ovarial transmission—the passage of WNV from female mosquitoes to larva through the ovary—may contribute to overwintering and early season amplification of WNV[10].

1. How would you include vertical transmission into the model structure?
2. How might the addition of trans-ovarial transmission affect disease transmission?

---

[9] Bugbee, L. and L. Forte. 2004. The discovery of West Nile Virus in overwintering Culex pipiens (Diptera: Culicidae) mosquitoes in Lehigh County, Pennsylvania. Journal of the American Mosquito Control Association. 20(3): 326–327.

[10] Baqar, S., C.G. Hayes, J.R. Murphy, and D.M. Watts. 1993. Vertical transmission of West Nile virus by Culex and Aedes species mosquitoes. American Journal of Tropical Medicine and Hygiene. 48: 757–762.

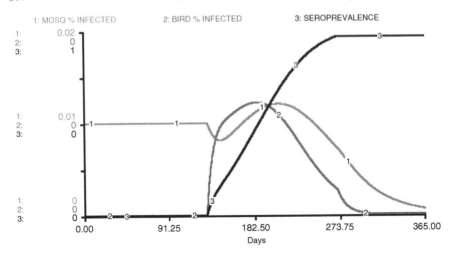

Fig. 12.7

3. What field data would help with estimating the rate of trans-ovarial transmission of WNV in mosquitoes? (Hint: Which sex does not bite?)

WNV BASE MODEL (DERIVED FROM)[6]

BIRD_INFECT(t) = BIRD_INFECT(t − dt) + (BIRD_INFECTING −
Bird_I_to_R − BI_DYING_WNV) * dt
INIT BIRD_INFECT = 0

INFLOWS:
BIRD_INFECTING = BITE_RATE*TRANSMISSION_EFFICACY_M_TO_B *
BIRD_SUSC/TOTAL_BIRD_POPULATION * MOSQ_INFECT
OUTFLOWS:
Bird_I_to_R = BIRD_INFECT * BIRD_RECOVERY_RATE
BI_DYING_WNV = BIRD_INFECT * WNV_DEATH_RATE
BIRD_RECOV(t) = BIRD_RECOV(t − dt) + (Bird_I_to_R −
IMMUNITY_LOSING) * dt
INIT BIRD_RECOV = 0

INFLOWS:
Bird_I_to_R = BIRD_INFECT * BIRD_RECOVERY_RATE
OUTFLOWS:
IMMUNITY_LOSING = BIRD_RECOV * IMMUNITY_LOSS_RATE
BIRD_SUSC(t) = BIRD_SUSC(t - dt) + (IMMUNITY_LOSING −
BIRD_INFECTING) * dt
INIT BIRD_SUSC = 1000

INFLOWS:
IMMUNITY_LOSING = BIRD_RECOV * IMMUNITY_LOSS_RATE

OUTFLOWS:
BIRD_INFECTING = BITE_RATE*TRANSMISSION_EFFICACY_M_TO_B *
BIRD_SUSC/TOTAL_BIRD_POPULATION * MOSQ_INFECT
MOSQ_EXPOS(t) = MOSQ_EXPOS(t − dt) + (MOSQ_INFECTING −
INCUBATION − ME_DYING) * dt
INIT MOSQ_EXPOS = 0

INFLOWS:
MOSQ_INFECTING = BITE_RATE*TRANSMISSION_EFFICACY_B_TO_M *
MOSQ_SUSCEP/TOTAL_ADULT_MOSQ*BIRD_INFECT
OUTFLOWS:
INCUBATION = MOSQ_EXPOS*TRANSITION_RATE
ME_DYING = MOSQ_EXPOS*MOSQ_DEATH_RATE
MOSQ_INFECT(t) = MOSQ_INFECT(t − dt) + (INCUBATION −
MI_DYING) * dt
INIT MOSQ_INFECT = 10000 * 0.01

INFLOWS:
INCUBATION = MOSQ_EXPOS * TRANSITION_RATE
OUTFLOWS:
MI_DYING = MOSQ_INFECT * MOSQ_DEATH_RATE
MOSQ_LARVA(t) = MOSQ_LARVA(t − dt) + (MOSQ_BIRTHING −
MOSQ_DEVELOPING − LARVAL_DYING) * dt
INIT MOSQ_LARVA = 0

INFLOWS:
MOSQ_BIRTHING = If time>130 then
OVIPOSITION_RATE*TOTAL_ADULT_MOSQ else 0
OUTFLOWS:
MOSQ_DEVELOPING = MOSQ_LARVA * MATURATION_RATE
LARVAL_DYING = MOSQ_LARVA * LARVAL_DEATH_RATE
MOSQ_SUSCEP(t) = MOSQ_SUSCEP(t − dt) + (MOSQ_DEVELOPING −
MOSQ_INFECTING − MS_DYING) * dt
INIT MOSQ_SUSCEP = 10000 − 10000 * 0.01

INFLOWS:
MOSQ_DEVELOPING = MOSQ_LARVA * MATURATION_RATE
OUTFLOWS:
MOSQ_INFECTING = BITE_RATE*TRANSMISSION_EFFICACY_B_TO_M*
MOSQ_SUSCEP/TOTAL_ADULT_MOSQ * BIRD_INFECT
MS_DYING = MOSQ_SUSCEP * MOSQ_DEATH_RATE
WNV_DEAD(t) = WNV_DEAD(t − dt) + (BI_DYING_WNV) * dt
INIT WNV_DEAD = 0

INFLOWS:
BI_DYING_WNV = BIRD_INFECT*WNV_DEATH_RATE
BIRD_%_INFECTED = Birds_Infected/ Total_Bird_Population

BIRD_%_RECOVERED = BIRD_RECOV/TOTAL_BIRD_POPULATION
BIRD_RECOVERY_RATE = 0.05
BITE_RATE = If time>150 then 0.20 else 0
IMMUNITY_LOSS_RATE = 0
LARVAL_DEATH_RATE = 1.191
MATURATION_RATE = 0.07
MOSQ_%_INFECTED = MOSQ_INFECT/TOTAL_ADULT_MOSQ
MOSQ_DEATH_RATE = If time>130 then 0.03 else 0.001
OVIPOSITION_RATE = 0.5
SEROPREVALENCE = (BIRD_INFECT + BIRD_RECOV)/
TOTAL_BIRD_POPULATION
TOTAL_ADULT_MOSQ = MOSQ_EXPOS + MOSQ_INFECT +
MOSQ_SUSCEP
TOTAL_BIRD_POPULATION = BIRD_SUSC + BIRD_RECOV +
BIRD_INFECT
TRANSITION_RATE = 0.1
TRANSMISSION_EFFICACY_B_TO_M = 0.2
TRANSMISSION_EFFICACY_M_TO_B = 0.9
WNV_DEATH_RATE = 0.05

12.4 Avian Population Effects and Seasonal Dynamics

12.4.1 Modifications to the Base Model

To further explore the role of host and vector population dynamics in this arboviral encephalitis system, we add model elements to incorporate host population growth and to more explicitly model seasonal dynamics. Figures 12.8 and 12.9 demonstrate the full model structure for the avian and mosquito components, respectively. Starting with the base model as described above, we incorporate a BIRD BIRTHING flow, which adds new individuals to the population through a simple exponential growth equation, and a background death rate with flows from each of the bird stocks, also modeled using the simple exponential population growth model. BIRTH RATE and DEATH RATE are conservatively estimated for a mixed bird community as 2.5 offspring per breeding season per individual (0.025 offspring per day with BIRD BIRTHING restricted to an estimated 3-month breeding season). BIRD DEATH RATE is set to produce similar peak population sizes each season for a 3-year simulation. MOSQUITO DEATH RATE is also calibrated to produce similar peak numbers during each simulated breeding season (Figure 12.10). This was done so that differences in disease prevalence in both the host and vector across the simulated seasons can be attributed to differences in disease transmission and not to major differences in peak populations. We address the role of inter-seasonal difference in population growth rates in disease transmission in section 12.3.

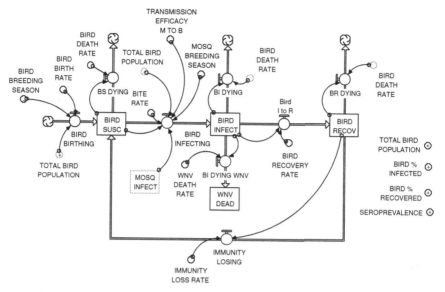

Fig. 12.8

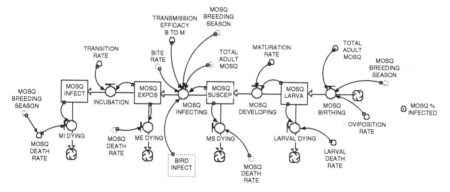

Fig. 12.9

Seasonal dynamics are imposed on the system using a simple graphical function for BIRD BREEDING SEASON, which modifies the rate of BIRD BIRTHING as a proportion of the maximum birth rate (0.025 offspring per day per individual) ranging from 0 to 1. MOSQUITO BREEDING SEASON is set similarly as a proportion changing over time to influence the rates of MOSQUITO BIRTHING, MOSQUITO EXPOSING, and BIRD INFECTING. These graphical functions allow for affected flows to operate between April and September for birds and between May and September for mosquitoes, a season typical of the central Midwestern United States.

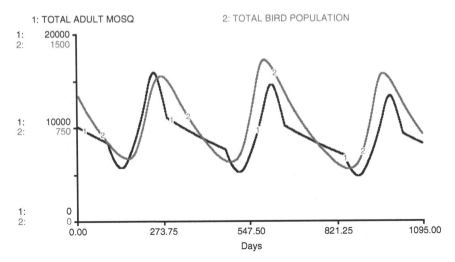

Fig. 12.10

12.4.2 Avian Demography and Disease Persistence

As we saw in our simple SIR model in section 12.3.1, a disease like WNV might easily sweep through a population rapidly and be extinguished without the addition of new susceptible individuals (Figures 12.7 and 12.8). Juvenile birds have been speculated to be key players in persistence of WNV in wild bird populations both due to their immunological naivety to the disease and due to an increased susceptibility to mosquito bites until they develop a full adult plumage[11]. If we run our expanded WNV model with the avian demographic rates (BIRD BIRTH RATE and BIRD DEATH RATE) set to zero, not surprisingly, we see a sharp rise in seroprevalence in the first year of exposure, with extinction of the disease due to lack of susceptible hosts in the second season (Figure 12.11). Alternatively, with seasonal avian population growth, we see seasonal reductions in seroprevalence due to the influx of new susceptible juveniles, followed by increasing eroprevalence as that season's outbreak progresses (Figure 12.12).

A sensitivity analysis of BIRD BIRTH RATE (with values ranging from 0.005 to 0.05 offspring per individual per day) demonstrates the effect of avian reproductive output on disease transmission. At the high end of the reproductive range (number 6 in Figures 12.13 and 12.14), seroprevalence in the host remains low while the infection prevalence in mosquitoes reaches high levels. At a low reproductive rate, seroprevalence in the avian hosts increases quickly resulting in low disease prevalence in the mosquito vector (Figures 12.13 and 12.14).

---

[11] Scott, T.W. and J.D. Edman 1991. Effects of avian host age and arbovirus infection on mosquito attraction and blood-feeding success. In: *Bird-Parasite Interactions: Ecology, Evolution, and Behavior.* Loye, J.E., and M. Zuk, eds. New York: Oxford University Press. 179–204.

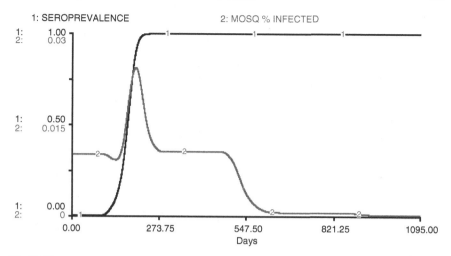

Fig. 12.11

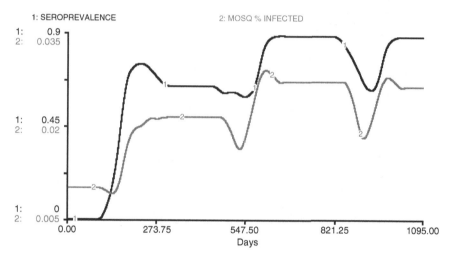

Fig. 12.12

12.4.3 Weather as an Extrinsic Driver of Outbreak Severity

Variations in seasonal weather patterns are a major driver of variation in WNV incidence among years[12]. This is primarily because mosquitoes require minimum temperatures to begin host seeking and oviposition. What happens if spring temperatures arrive earlier or later than average, changing the timing and duration

---

[12] Andreadis, T.G., J.F. Anderson, C.R. Vossbrinck, and A.J. Main. 2004. Epidemiology of West Nile virus in Connecticut: A five-year analysis of mosquito data 1999–2003. Vector-Borne and Zoonotic Disease. 4: 360–378.

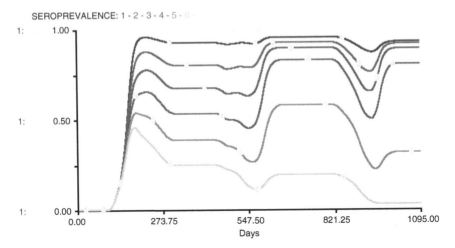

Fig. 12.13

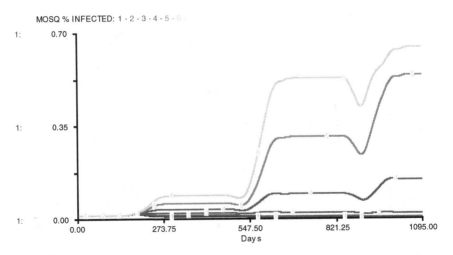

Fig. 12.14

of mosquito activity? To investigate this question, we shifted the MOSQUITO
BREEDING SEASON graph to begin one month earlier (Line 2 in Figures 12.15
and 12.16) or to begin one month later (Line 3 in Figures 12.15 and 12.16). Line
1 in these figures is the base seasonal onset of activity (from approximately April
to September) as was used in the previous section. What we see is that an earlier
onset of mosquito activity results in a longer period during which to amplify the
disease through the host-vector feedback loop. This affects the host as a higher cumulative exposure (peak SEROPREVALENCE) for birds which occurs earlier in

SEROPREVALENCE: 1 - 2 - 3 -

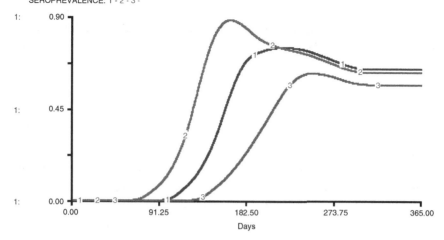

Fig. 12.15

TOTAL ADULT MOSQ: 1 - 2 - 3 -

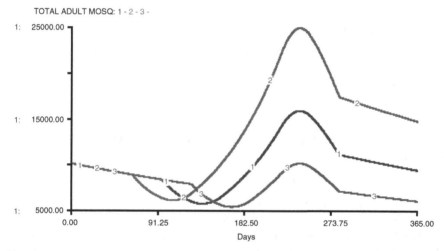

Fig. 12.16

the season (Figure 12.15, Line 2). This occurs because of the earlier onset of exponential growth for the mosquito population (Figure 12.16, Line 2) resulting in high mosquito numbers during the peak time of disease transmission.

The ecology of vector-borne, multi-host diseases like WNV poses a true challenge for modelers to balance model transparency and simplicity with the many layers of interaction that contribute to disease persistence and transmission. But these efforts have great potential to help public health officials, researchers, and policy makers better manage current disease concerns and prepare for the next emerging pathogen. The models presented here, as is often the case, are but a rudimentary attempt to simulate the dynamics of vastly more complicated disease ecology.

However, even in this relatively crude state, these models are useful tools that can offer great insight and help in formulating hypotheses about the behavior of disease outbreaks in the real world.

12.4.4 Questions and Tasks

[5] describe a phenomenon where species diversity may serve to reduce disease risk to humans. Some studies suggest that the number of human cases of WNV is decreased in areas of high bird diversity because the bird hosts, which develop the highestviremia and transmit the disease most efficiently when bitten by a mosquito, are a lower relative proportion of the total population.

1. How would you explore the potential effects of avian community diversity on human risk of disease in this model? (Hint: Try a sensitivity analysis and graph MOSQUITO % INFECTED).

WNV EXPANDED SEASONAL MODEL

BIRD_INFECT(t) = BIRD_INFECT(t − dt) + (BIRD_INFECTING − Bird_I_to_R − BI_DYING − BI_DYING_WNV) * dt
INIT BIRD_INFECT = 0

INFLOWS:
BIRD_INFECTING =
BITE_RATE*TRANSMISSION_EFFICACY_M_TO_B*BIRD_SUSC/
TOTAL_BIRD_POPULATION * MOSQ_INFECT *
MOSQ_BREEDING_SEASON
OUTFLOWS:
Bird_I_to_R = BIRD_INFECT * BIRD_RECOVERY_RATE
BI_DYING = BIRD_INFECT * BIRD_DEATH_RATE
BI_DYING_WNV = BIRD_INFECT * WNV_DEATH_RATE
BIRD_RECOV(t) = BIRD_RECOV(t − dt) + (Bird_I_to_R −
IMMUNITY_LOSING − BR_DYING) * dt
INIT BIRD_RECOV = 0

INFLOWS:
Bird_I_to_R = BIRD_INFECT*BIRD_RECOVERY_RATE
OUTFLOWS:
IMMUNITY_LOSING = BIRD_RECOV*IMMUNITY_LOSS_RATE
BR_DYING = BIRD_RECOV*BIRD_DEATH_RATE
BIRD_SUSC(t) = BIRD_SUSC(t − dt) + (IMMUNITY_LOSING +
BIRD_BIRTHING − BIRD_INFECTING − BS_DYING) * dt
INIT BIRD_SUSC = 1000

INFLOWS:
IMMUNITY_LOSING = BIRD_RECOV*IMMUNITY_LOSS_RATE

BIRD_BIRTHING =
BIRD_BIRTH_RATE*TOTAL_BIRD_POPULATION*
BIRD_BREEDING_SEASON
OUTFLOWS:
BIRD_INFECTING =
BITE_RATE*TRANSMISSION_EFFICACY_M_TO_B*
BIRD_SUSC/TOTAL_BIRD_POPULATION*MOSQ_INFECT*
MOSQ_BREEDING_SEASON
BS_DYING = BIRD_SUSC*BIRD_DEATH_RATE
MOSQ_EXPOS(t) = MOSQ_EXPOS(t − dt) + (MOSQ_INFECTING −
INCUBATION − ME_DYING) * dt
INIT MOSQ_EXPOS = 0

INFLOWS:
MOSQ_INFECTING = BITE_RATE*TRANSMISSION_EFFICACY_B_TO_M*
MOSQ_SUSCEP/TOTAL_ADULT_MOSQ * BIRD_INFECT *
MOSQ_BREEDING_SEASON
OUTFLOWS:
INCUBATION = MOSQ_EXPOS * TRANSITION_RATE
ME_DYING = MOSQ_EXPOS * MOSQ_DEATH_RATE
MOSQ_INFECT(t) = MOSQ_INFECT(t − dt) + (INCUBATION −
MI_DYING) * dt
INIT MOSQ_INFECT = 10000*0.01

INFLOWS:
INCUBATION = MOSQ_EXPOS * TRANSITION_RATE
OUTFLOWS:
MI_DYING = MOSQ_INFECT * MOSQ_DEATH_RATE
MOSQ_LARVA(t) = MOSQ_LARVA(t − dt) + (MOSQ_BIRTHING −
MOSQ_DEVELOPING − LARVAL_DYING) * dt
INIT MOSQ_LARVA = 0

INFLOWS:
MOSQ_BIRTHING = OVIPOSITION_RATE*TOTAL_ADULT_MOSQ*
MOSQ_BREEDING_SEASON
OUTFLOWS:
MOSQ_DEVELOPING = MOSQ_LARVA * MATURATION_RATE
LARVAL_DYING = MOSQ_LARVA * LARVAL_DEATH_RATE
MOSQ_SUSCEP(t) = MOSQ_SUSCEP(t − dt) + (MOSQ_DEVELOPING −
MOSQ_INFECTING − MS_DYING) * dt
INIT MOSQ_SUSCEP = 10000 − 10000*0.01

INFLOWS:
MOSQ_DEVELOPING = MOSQ_LARVA * MATURATION_RATE
OUTFLOWS:
MOSQ_INFECTING = BITE_RATE*TRANSMISSION_EFFICACY_B_TO_M*
MOSQ_SUSCEP/TOTAL_ADULT_MOSQ * BIRD_INFECT*

MOSQ_BREEDING_SEASON
MS_DYING = MOSQ_SUSCEP * MOSQ_DEATH_RATE
WNV_DEAD(t) = WNV_DEAD(t − dt) + (BI_DYING_WNV) * dt
INIT WNV_DEAD = 0

INFLOWS:
BI_DYING_WNV = BIRD_INFECT * WNV_DEATH_RATE
BIRD_%_INFECTED = Birds_Infected/ Total_Bird_Population
BIRD_%_RECOVERED = BIRD_RECOV/TOTAL_BIRD_POPULATION
BIRD_BIRTH_RATE = 0.025
BIRD_DEATH_RATE = 0.005
BIRD_RECOVERY_RATE = 0.10
BITE_RATE = 0.3
IMMUNITY_LOSS_RATE = 0
LARVAL_DEATH_RATE = 1.191
MATURATION_RATE = 0.07
MOSQ_%_INFECTED = MOSQ_INFECT/TOTAL_ADULT_MOSQ
MOSQ_DEATH_RATE = If MOSQ_BREEDING_SEASON>0 then 0.015
else 0.002
OVIPOSITION_RATE = 0.5
SEROPREVALENCE = (BIRD_INFECT + BIRD_RECOV)/
TOTAL_BIRD_POPULATION
TOTAL_ADULT_MOSQ = MOSQ_EXPOS + MOSQ_INFECT +
MOSQ_SUSCEP
TOTAL_BIRD_POPULATION = BIRD_SUSC + BIRD_RECOV +
BIRD_INFECT
TRANSITION_RATE = 0.1
TRANSMISSION_EFFICACY_B_TO_M = 0.3
TRANSMISSION_EFFICACY_M_TO_B = 0.9
WNV_DEATH_RATE = 0.05
BIRD_BREEDING_SEASON = GRAPH(TIME)
(0.00, 0.00), (31.3, 0.00), (62.6, 0.00), (93.9, 0.00), (125, 0.1), (156, 0.4), (188,
1.00), (219, 1.00), (250, 0.4), (282, 0.1), (313, 0.00), (344, 0.00), (375, 0.00),
(407, 0.00), (438, 0.00), (469, 0.1), (501, 0.4), (532, 1.00), (563, 1.00), (594, 0.4),
(626, 0.1), (657, 0.00), (688, 0.00), (720, 0.00), (751, 0.00), (782, 0.00), (813,
0.00), (845, 0.1), (876, 0.4), (907, 1.00), (939, 1.00), (970, 0.4), (1001, 0.1),
(1032, 0.00), (1064, 0.00), (1095, 0.00)
MOSQ_BREEDING_SEASON = GRAPH(TIME)
(0.00, 0.00), (31.3, 0.00), (62.6, 0.00), (93.9, 0.00), (125, 0.3), (156, 1.00), (188,
1.00), (219, 1.00), (250, 0.3), (282, 0.00), (313, 0.00), (344, 0.00), (375, 0.00),
(407, 0.00), (438, 0.00), (469, 0.00), (501, 0.3), (532, 1.00), (563, 1.00), (594,
1.00), (626, 0.3), (657, 0.00), (688, 0.00), (720, 0.00), (751, 0.00), (782, 0.00),
(813, 0.00), (845, 0.00), (876, 0.3), (907, 1.00), (939, 1.00), (970, 1.00), (1001,
0.3), (1032, 0.00), (1064, 0.00), (1095, 0.00)

Chapter 13
Chaos and Pestilence

Before the advent of modern computer technology and software, many modeling efforts and scientific experiments were designed for linear, often static systems, which had the advantage of being analytically solvable. The ways of thinking about system behavior and the tools applied to describe that behavior was rooted deeply in classical mechanics. This science was used to describe the behavior of whole classes of moving objects, such as pendulums, falling rocks, or projectiles. The scientific paradigms associated with classical mechanics were not only applied in the realm of the natural sciences but increasingly influenced models of economic and ecological systems as well.

The strength of these paradigms lies in their view of systems as predictable, well-described entities that can be analyzed with available mathematical tools. Students were told that nonlinear systems are generally unsolvable and that such systems are exceptions. The first of these statements is true; nonlinear systems, some of which we modeled in the previous chapters, generally do not have an explicit mathematical solution. However, the second statement, that nonlinear systems are exceptions, is false. Rather, many real systems are governed by nonlinearities. These systems frequently exhibit characteristics that were previously unanticipated or misidentified.

The emergence of chaos theory made us aware of the importance of nonlinearities, a lack of predictability that is inherent in many of these nonlinear systems, the sensitivity of model results to small changes in initial conditions, and therefore, the need for increased computer modeling efforts. Today, chaos theory begins to influence thinking in modern natural sciences as well as in the social sciences. In the following sections of this chapter, we develop models with potentially chaotic behavior first in the context of the spread of a disease—akin to the simple models in Chapter 2—and then in the context of insect dynamics and associated host–parasitoid interactions, which we touch on throughout the book.

B. Hannon and M. Ruth, *Dynamic Modeling of Diseases and Pests*,
Modeling Dynamic Systems,
© Springer Science+Business Media LLC 2009

13.1 Basic Disease Model with Chaos

13.1.1 Model Set-up

Assume the size of the sick population affects its awareness about a disease (Figure 13.1) and thus its behaviors and rate at which individuals in the population become contagious (Figure 13.2). GETTING SICK is the product of the CONTAGION RATE and the stock of SICK people. GETTING WELL, similarly, is a function of a "sick time coefficient" and the stock. However, to allow for the emergence of chaos, we have specified the relationship as a nonlinear equation:

$$\text{GETTING WELL} = \text{SICK}^\wedge 2 / \text{SICK\_TIME\_COEFF} \qquad (13.1)$$

Running the model with a CONTAGION RATE as in Figure 13.2, but setting the maximum value to .2 instead of 2 yields the results shown in Figure 13.3. The number of sick initially increases and comes to a steady state. Changing the maximum CONTAGION RATE to 1.2 can lead to an initial overshoot and ultimate steady state (Figure 13.4). Increasing the CONTAGION RATE maximum further (and/or giving your curve of Figure 13.2 more curvature) can yield periodic oscillations (Figure 13.5) and ultimately chaos (Figure 13.6). The results in Figure 13.6 are derived for the relationship between AWARENESS LEVEL and CONTAGION RATE of Figure 13.2.

While the chaotic time path of Figure 13.6 does not show any order—no pattern repetitions are seen in the behavior of our state variable—a neat relationship exists nonetheless between the values of the state variable at any given period of time and the value it took on a DT earlier. This relationship is shown in Figure 13.7, and emergence of the pattern depicted there can be used as a heuristic means when trying to detect chaos.

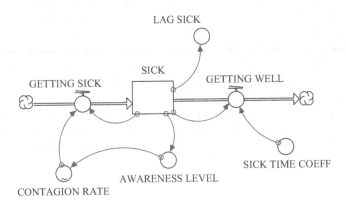

Fig. 13.1

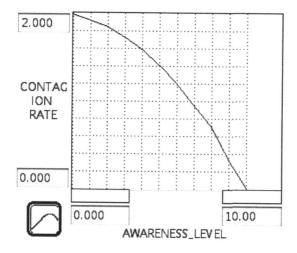

Fig. 13.2

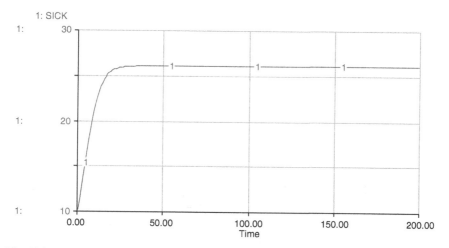

Fig. 13.3

13.1.2 Detecting and Interpreting Chaos

The only way to technically determine the presence of chaos in our model is to be able to calculate the so-called Lyapunov exponent for the model. If this exponent is greater than zero, the model is chaotic. The exponent can be calculated using data from the model on the rate of separation of nearest neighboring points though time. It seems that at our level of modeling, however, we can observe the model results.

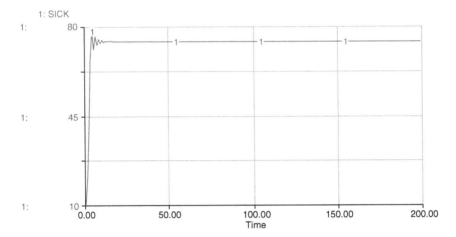

Fig. 13.4

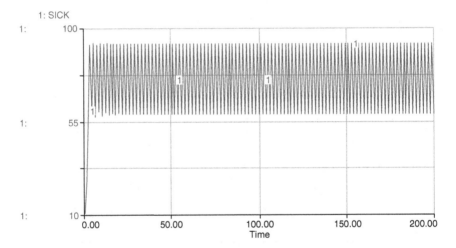

Fig. 13.5

Sensitivity of the nonlinear model to initial conditions is an attribute of chaotic models. Models that show chaos may not show it for all combinations of the parameters and detecting the advent of chaos becomes the issue. In these cases, we can examine a scatter plot of the state against the one-DT lagged value of that state. If we see the formation of a pattern, such as a curve, where the successive points are landing on opposite sides of that curve (each time in a different place), we have a model with chaos. If this graph show oscillation between opposite sides of the curve but always between the same two points, we have a model in oscillation. If the curve shows progression to a point, we have a model at steady state. Some models will start off oscillating and then dampen down to a steady state. These descriptions

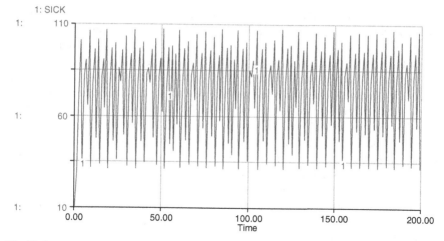

Fig. 13.6

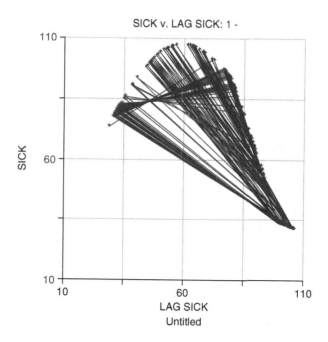

Fig. 13.7

of the appearance of chaos in models can be helpful to the beginning modeler. The
key idea is to recognize when one has chaos and when one has random variation.
In a time series,plot both will appear to be random. However, in the scatter plot, the
model's randomness appears as a uniform randomness across the range of the plot.
If random variation appears in a time series plot of a model containing no random

functions, the model is either in chaos or (more likely), the choice of DT is too large. The DT choice in the latter case is a longer time than the critical time of the model. Shortening the DT will cause such a model to behave with at least some smooth curves. For the most part, these smooth curves will have continuous derivatives. Parameter change can bring a return of chaotic behavior, and in the nonchaotic model, bring back of return of random behavior. The best test here is an examination of the model in the scatter plot and a check on the sensitivity to initial condition changes.

At this point we should ask if chaos occurs in nature. We find that, indeed, it does. Chaos is evident in the variations of heartbeats and brainwaves or the irregularity of water dripping from a faucet. Both living and nonliving systems seem to show chaos. Why? To what advantage is such a result to these systems? Stuart Kaufman,[1] among others, has proposed that all systems seem to evolve toward higher and higher efficiencies of operation. Many systems are so highly disturbed by variations in their environment that their efficiencies are rarely high. However, if these disturbances can be held to a minimum, then the evolution of the system becomes more complete and more efficient but closer to the border of chaotic behavior.

Earthquakes and avalanches are examples of energy-storing systems that continuously redistribute the incoming stresses more and more efficiently until a breaking point is reached and the border to chaos is opened. Does this mean that the brain and the heart have somehow evolved closer to some maximum efficiency for such organisms? We do not know the answer. We do know that the scale of measurement matters. For example, if we were to watch the pattern on a patch of natural forest over many centuries, we would see the rise and sharp fall of the biomass levels, unpredictably. Forest fires and insects find ample host in such forest patches once they have developed a large amount of dry biomass bound up in relatively few species. The patch evolves or succeeds to greater and greater efficiency of light energy conversion by getting larger and fewer species. But the patch also becomes more vulnerable to fire and pests, and eventually collapses. Yet if we look at the total biomass on a large collection of such biomasses, whose collapses are not synchronized, this total biomass remains relatively constant. Thus, chaotic-like behavior in the small is not seen in the large. Could this mean that natural systems have "found" chaos in their search for greater efficiencies and have "learned" to stagger the chaotic events, allowing faster rebound and large-scale stability? We do not know the answers, but we think the implications are fascinating.

13.1.3 Questions and Tasks

1. Re-create the model described in this chapter, but try to find chaos in a different way. Instead of increasing the maximum value for the CONTAGION RATE for subsequent runs in the graph of Figure 13.2, set that maximum to 1.4 and increase the curvature of the graph. Can you make chaos occur? What does this mean for the underlying mechanisms behind this epidemic?

---

[1] Kaufman, S. 1993. The Origins of Order: Self-Organization and Selection in Evolution, New York, Oxford University Press.

2. Introduce a vaccination program into your chaotic epidemic model. How would the program need to be structured to prevent "chaos" (in the modeling sense)?
3. In the model, can you find chaos for a fixed (constant) contagion rate simply by varying SICK TIME COEFF? Interpret your results.
4. Change the exponent in the GETTING WELL equation to values

   a. less than 2
   b. larger than 2.

   Interpret your findings.

BASIC EPIDEMIC MODEL WITH CHAOS

SICK(t) = SICK(t − dt) + (GETTING_SICK − GETTING_WELL) * dt
INIT SICK = 10

INFLOWS:
GETTING_SICK = CONTAGION_RATE * SICK
OUTFLOWS:
GETTING_WELL = SICK^2/SICK_TIME_COEFF
AWARENESS_LEVEL = .1 * SICK
LAG_SICK = DELAY(SICK,DT)
SICK_TIME_COEFF = 150
CONTAGION_RATE = GRAPH(AWARENESS_LEVEL)
(0.00, 1.99), (1.00, 1.92), (2.00, 1.84), (3.00, 1.73), (4.00, 1.58), (5.00, 1.41), (6.00, 1.19), (7.00, 0.94), (8.00, 0.7), (9.00, 0.32), (10.0, 0.00)

13.2 Chaos with Nicholson-Bailey Equations[2]

13.2.1 Host-Parasitoid Interactions

In Chapter 3, we modeled the spread of a parasitic infection in an insect population of two life stages. The focus of that model was the spread of the infection. Therefore, we ignored the fate of the parasitoid. In this chapter however, we model explicitly the interactions between the host and the parasitoid populations. Rather than setting up our model in terms of population sizes, we specify host–parasitoid interactions in terms of population densities.

In order to model the host–parasitoid interactions, we abstract away from the fact that only specific life cycle stages exhibit those interactions. After you worked through this chapter, you may want to refine the model to account for the fact that,

---

[2] This chapter follows L. Ederstein-Keshet, Mathematical Models in Biology, Random House, 1988, 79–85, and D. Brown and P. Rothery, Models in Biology: Mathematics, Statistics and Computing, Wiley, NY, 1993, pp. 399–406.

for example, adult parasitoids lay their eggs in the pupae of hosts, but not in the eggs of their hosts or with the larvae or adults.

Denote, respectively, H(t) and P(t) as the host and parasitoid densities in time period t, and F(H(t), P(t)) as the fraction of hosts that is not parasitized. Then

$$H(t+1) = \lambda * H(t) * F(H^t), P(t)) \tag{13.2}$$

$$P(t+1) = C * H(t) * [1 - F(H(t), P(t))] \tag{13.3}$$

where $\lambda(H(t))$ is the host growth rate and C is the parasitoid fecundity.

Let us assume that the fraction of hosts that become parasitized depends on the density-dependent rate of encounter of parasitoids and hosts. Encounters occur randomly, allowing us to invoke the law of mass action that we used extensively throughout this book to model the spread of disease from contagious to susceptible populations. Accordingly, the number of encounters of hosts HE with parasitoids is

$$HE(t) = A * H(t) * P(t) \tag{13.4}$$

where A is the searching efficiency of the parasitoids.

Unlike the models of the spread of a disease from an infected to a nonimmune population, subsequent encounters of individuals in the two populations do not alter the rate at which parasitoids are propagated. Therefore, we need to modify the law of mass action to account for the fact that only the first encounter of hosts and parasitoids is significant in propagating the parasitoid. Once a host carries the parasitoid's eggs, subsequent encounters with parasitoids will not change the number of parasitoid progeny that hatch from the host. We need only to distinguish between hosts that had no encounter and hosts that had at least one encounter with parasitoids.

The Poisson distribution describes the occurrence of such discrete, random events as encounters of hosts and parasitoids. We can make use of the Poisson probability distribution to calculate the probability that there is no attack of parasitoids on a host within a certain time period. In general, therefore

$$P(X) = \frac{\text{EXP}\left(-\frac{HE(t)}{H(t)}\right)\left(\frac{HE(t)}{H(t)}\right)^X}{X!} \tag{13.5}$$

is the probability of X attacks. This probability depends on the average number of attacks in the given time interval, HE/H. From equation (13.4) we know

$$HE(t)/P(t) = A*P(t) \tag{13.6}$$

Thus, for zero attacks by the parasitoids, equation (13.5) yields

$$P(0) = \frac{\text{EXP}(-A*P(t))(A*P(t))^0}{0!} = \frac{\text{EXP}(-A*P(t))*1}{1} = \text{EXP}(-A*P(t))$$

$$(13.7)$$

Equations (13.2) and (13.3) can therefore be re-written as

$$H(t+1) = H(t)*\lambda*\text{EXP}(-A*P(t))$$

$$(13.8)$$

$$P(t+1) = C*H(t)*[1 - \text{EXP}(-A*P(t))]$$

$$(13.9)$$

Let us also assume that without parasitoids, the hosts will grow toward a carrying capacity K set by the environment. To capture growth of the host population up to a density $H(t) = K$ and decline of the host population for $H(t) > K$, we replace in equation (13.8) the growth rate $\lambda(H(t))$ with

$$\lambda = \text{EXP}\left(R*\left(1 - \frac{H(t)}{K}\right)\right)$$

$$(13.10)$$

where R is the maximum host growth rate. Thus, the equation governing the size of the host population in time t+1 becomes

$$H(t+1) = H(t)*\text{EXP}\left(R*\left(1 - \frac{H(t)}{K}\right) - A*P(t)\right)$$

$$(13.11)$$

and after subtracting the respective state variables in time period t from equations (13.9) and (13.11), we have a set of differential equations that capture the change of host and parasitoid densities from time period t to $t + 1$:

$$\Delta H(t) = H(t)*\text{EXP}\left(R*\left(1 - \frac{H(t)}{K}\right) - A*P(t)\right) - H(t)$$

$$(13.12)$$

$$\Delta P(t) = C*H(t)*[1 - \text{EXP}(-A*P(t))] - P(t)$$

$$(13.13)$$

We can now see the dynamics exhibited by this model (Figure 13.8). These equations describing changes in the host and parasitoid densities can yield a variety of results, from the production of steady state conditions for the host and parasitoid, to their lock in a limit cycle, to chaos.

The following graphs result from the parameters and initial conditions in the table, and a DT = 1:

13.2.2 Questions and Tasks

1. We have modeled in this section of Chapter 13 one type of species interaction that is almost exclusively found among insects. Typically, both the parasitoid and host

Table 13.1

Figure	Description	R	A	K	H(t = 0)	P(t = 0)
13.9	Steady State	0.50	0.20	14.5	10.00	1.00
13.10	Limit Cycle	2.00	0.20	21.5	10.00	1.00
13.11	Chaos	2.65	0.20	25.0	10.00	1.00

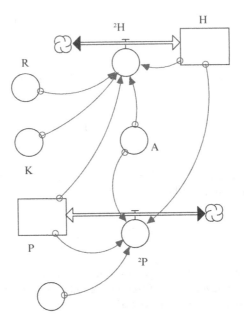

Fig. 13.8

have a number of life cycle stages—eggs, larvae, pupae, and adults—and their interaction is limited to a subset of these.

a. Can you modify the model to account for the fact that it is typically only the larvae of the host that get parasitized by adult parasitoids?
b. How does this disaggregation of the parasitoid and host population affect your results?
c. Can you find parameters and initial values that generate alternatively steady state, limit cycles, or chaos?
d. What is the appropriate DT to use here and how are the results affected by its choice?

2. Try reducing the DT. In the earlier graphs, DT is set at 1.00. A smaller DT yields a completely different answer. What is going on here? Is the DT of 1.00 required by the host or the parasitoid? (The Nicholson-Bailey model views t = 1 as one generation and all the dynamics for one DT go on inside that time period of 1.00.

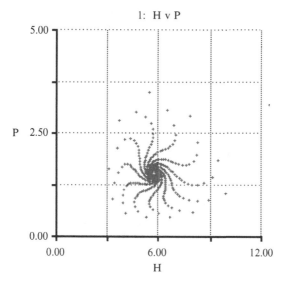

Fig. 13.9

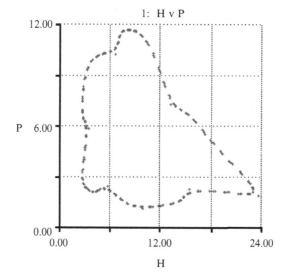

Fig. 13.10

It is as though the whole new generation of the two populations is formed just before the beginning of that generation. So in this model, a DT less than 1.00 has no meaning.)
3. Do a sensitivity analysis on the initial values of H and P, on R, A, and K.

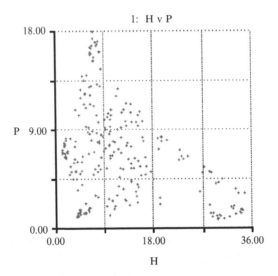

Fig. 13.11

NICHOLSON-BAILEY HOST–PARASITOID MODEL

H(t) = H(t − dt) + (ΔH) * dt
INIT H = 10

INFLOWS:
ΔH = H * EXP(R * (1 − H/K) − A * P) − H
P(t) = P(t − dt) + (ΔP) * dt
INIT P = 1

INFLOWS:
ΔP = C * H * (1 − EXP(−A * P)) − P
A = .2
C = 1
K = 14.5
R = .5

Chapter 14
Catastrophe and Pestilence

14.1 Basic Catastrophe Model

If a large number of real systems exhibit dynamics that bear the potential for chaos, why do we not see more chaos in real-world processes? Fortunately, the domains over which stability of the system occurs can be relatively large. But once in a while, systems may move "toward the edge of stability" and little nudges to the system may move it from stability to instability—that is, into a catastrophe. Subsequently, reorganization of system components may occur to bring the system back into a stable domain—a kind of evolutionary process. This stable domain, however, may not be the same as the one prior to the disturbance.

The system undergoes a catastrophic event in the sense that it is moved from an initial state of stability through a dramatic phase of reorganization and back to some degree of stability. Examples for such catastrophic events include landslides, avalanches, earthquakes, and pest outbreaks in ecosystems. In each case, small changes in the system occur that individually may not be critical to the system's behavior. Collectively, however, they lead to the evolution of the system toward a critical state. This is apparent, for example, in the case of avalanches. Each individual snowflake potentially adds to the instability of the system. When a critical point is reached, the next snowflake may trigger an avalanche that affects a large part of the system. Temporary stability is quickly reached if the avalanche is not too dramatic. Even if not of a large scale, the avalanche adds to the "stress" of the system downhill, making it more susceptible to further avalanches as more snow falls at those regions or as additional small avalanches are received from higher on the hill. Ultimately, a large-scale, catastrophic event may occur, which affects the entire system, not just individual regions. The system components re-group and finally enter a phase of new, temporary stability.

So, evolutionary processes are at work making the system more "efficient" in some sense. This is evolution toward catastrophe. A system in such a state can re-merge to a stable state by another process of evolution, likely faster than the first kind, and this new stable state may be inefficient. Large living natural systems are

B. Hannon and M. Ruth, *Dynamic Modeling of Diseases and Pests*,
Modeling Dynamic Systems,
© Springer Science + Business Media LLC 2009

likely constrained from operating at or near peak efficiency by random intervention of uncoordinated external processes at the regional levels.

In this chapter, we develop first a simple model of catastrophe and return then to modeling pest outbreaks. Let us start by considering the following figure[1] that illustrates the surface defined by the following equation:

$$X^3 - \text{ALPHA} * X - \text{BETA} = 0 \qquad\qquad (14.1)$$

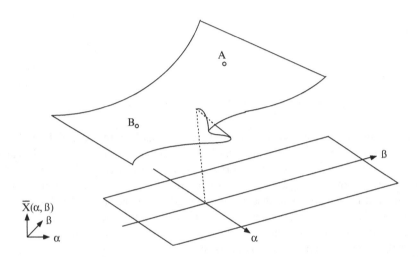

Imagine a stationary ball placed at the top of this surface, at point A. Small nudges away from its equilibrium point A will lead to a new equilibrium. After a series of such small perturbations, however, the ball will roll off the top part of the surface, and a priori, it is difficult for us to determine exactly where it will end up. All we know for certain is that the new equilibrium position is somewhere at the bottom of the surface, say point B.

Small nudges to the ball in B will again move it slightly away from B. And if we kick it hard enough, we can propel the ball through the fold, or "cusp," to the upper part of the surface again. Where exactly will it end up? To give a precise answer requires exact knowledge of the shape of the surface, the properties of the ball, and the magnitude and direction of the force exerted on the ball. In more complicated, real-life systems, some variables to describe the system and the forces incident upon them are unknown. As a result, we may only know stability domains rather than specific locations.

The STELLA model for equation (14.1) is given in Figure 14.1. We slightly vary X with each simulation time step. Solve equation (14.1) for BETA. Set DT = .0025 and define X, for example, as

---

[1] See Beltrami, E. 1987. *Mathematics for Dynamic Modeling*, Academic Press, Inc., Boston.

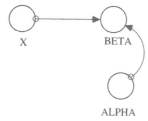

Fig. 14.1

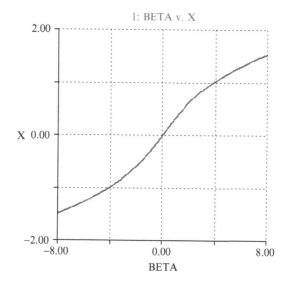

Fig. 14.2

$$X = \text{TIME}-2, \qquad\qquad (14.2)$$

then run the model.

For positive ALPHA, the cusp or fold appears in the X vs. BETA plot. For ALPHA = 0 and negative ALPHA, the "S" curve appears. The cases for negative, zero and positive values of ALPHA are shown in Figures 14.2 to 14.4, respectively. For illustration, we chose ALPHA = -3 in Figure 14.2 and ALPHA = $+3$ in Figure 14.4.

BASIC CATASTROPHE MODEL
ALPHA = -3 BETA = X^3—ALPHA * X   X = TIME—2

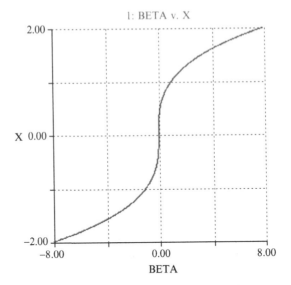

Fig. 14.3

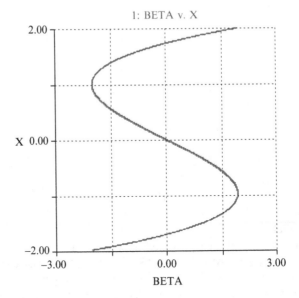

Fig. 14.4

14.2 Spruce Budworm Catastrophe

A classical example for the implications of catastrophes for ecosystem management is the spruce budworm dynamics. Spruce budworm is a caterpillar that feeds
on spruce and fir forests in the northeastern United States and eastern Canada. For

many years, population sizes of spruce budworm remain low and have little impact on trees. When forest stands reach maturity, however, spruce budworm populations explode, seriously affecting the forest by defoliating the trees. As a result of defoliation, trees are weakened and ultimately may die. With the death of trees comes a loss of the food source for spruce budworm and a consequent population crash.

The cycle of low spruce budworm population densities, followed by population explosions and catastrophic collapse tends to repeat over the course of years. The resulting damage and death of trees negatively affects the timber and paper pulp industries of the region. Frequently, forest managers decided to spray forest stands to control budworm populations. The dynamics inherent in the system, however, lead it to follow its own path, making ever more extensive pest control necessary. If those controls fail, outbreaks will be more severe and devastating than if the system had been left to control itself, as recent experiences in the United States and Canada show.

One of the most unused natural system controls in forestry relates to the idea of patch size. Natural systems no doubt avoid large catastrophes because they operate in patches, where the degree of maturity of adjacent patches is nearly always different. Consequently, pests and fires find difficulty in spreading beyond a patch, and the size of the catastrophe is kept small. Current forestry practice seems to be disconnected from such natural system behavior.

To model the spruce budworm catastrophe[2] let us denote B as the budworm population size, K as the carrying capacity, S as habitat size, and GR as the budworm's natural rate of increase. Thus,

$$\frac{dB}{dt} = GR * B * \left(1 - \frac{B}{K * S}\right) \tag{14.3}$$

would describe the population dynamics for a fixed carrying capacity and no predatory influences on population growth. This is the logistic growth equation that we have used in this book many times before. Let us introduce the effects of predation with a maximum predation rate C, which is assumed to be constant. At small population densities, predation has an insignificant effect on the budworm population because they are well hidden in a relatively dense canopy. As population densities increase, however, predators may increasingly feed on budworm that partially or totally defoliated the trees and are then easy prey. A predation term that captures such interactions is

$$\frac{C * B^2}{A^2 + B^2} \tag{14.4}$$

with A as a scalar that captures the effectiveness of the predators to spot and prey on spruce budworm. In an immature forest, predation is easier than in a mature forest with a diverse and dense canopy. Thus, A may be assumed to increase with increased maturity of the forest, i.e. habitat size S

---

[2] This model follows closely the model laid out in Beltrami, E. 1987. *Mathematics for Dynamic Modeling*, Academic Press, Inc., Boston, pp. 189—196.

$$A = K1 * S \tag{14.5}$$

and thus

$$\frac{C*B^2}{A^2+B^2} = \frac{C*B^2}{(K_1*S)^2+B^2}, \tag{14.6}$$

with K1 as a constant.

Combining predation with the logistic growth function yields

$$\Delta B = \frac{dB}{dt} = GR*B*\left(1-\frac{B}{K*S}\right) - \frac{C*B^2}{(K_1*S)^2+B^2} \tag{14.7}$$

which is the equation used in the model to drive spruce budworm population changes, ΔB.

Changes in habitat size are assumed to also follow the logistic growth curve, with RS as the natural rates of increase and KS as carrying capacity:

$$\Delta S = \frac{dS}{dt} = RS*S*\left(1-\frac{S}{KS*E}\right) \tag{14.8}$$

E is the percentage of foliage on trees. The more healthy the forest, the higher E. The percentage of foliage on trees is assumed to decrease as the average budworm density per habitat size B/S increases. To model diminishing stress as budworm populations decrease, we multiply B/S by E^2. The combined effect of logistic growth in foliage and budworm-induced foliage losses is

$$\Delta E = \frac{dE}{dt} = RE*E*\left(-\frac{P*B*E^2}{S}\right) \tag{14.9}$$

with RE the rate of foliage increase and P a proportionality factor.

Let us consider the case of $B \neq 0$ and introduce the following notation:

$$R' = \frac{R*K1*S}{B} \tag{14.10}$$

$$Q = \frac{K}{K1} \tag{14.11}$$

and rewrite

$$B = K1*S*X \tag{14.12}$$

It can be shown[3] that the nontrivial equilibria of equation (14.7) fulfill

$$R'\left(1-\frac{X}{Q}\right) = G(X) \tag{14.13}$$

with

$$G(X) = \frac{X}{1+X^2} \tag{14.14}$$

---

[3] Beltrami, E. 1987. *Mathematics for Dynamic Modeling*, Academic Press, Inc., Boston.

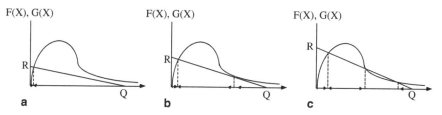

Fig. 14.5

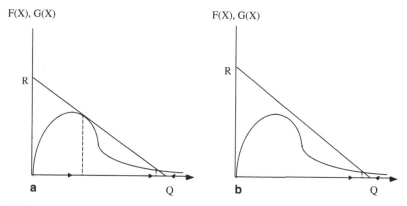

Fig. 14.6

The left side of equation (14.13) is a straight line F(X) with slope $-R'/Q$. Equilibria occur where this line intersects with G(X).

S and R increase with increases in Q. At first, there is a single equilibrium, corresponding to the situation shown in Figure 14.5(a). After some time, the line becomes tangent to the curve, as shown in Figure 14.5(b). With further increases in the slope, two points that "attract" system behavior emerge (Figure 14.5(c))—these are the two outer points.

As the slope of R increases even further, another point of tangency is realized, (Figure 14.6(a)) and from thereon, only one point of intersection (Figure 14.6(b)), a stable attractor, persists:

The cusp of the spruce budworm dynamics is shown schematically in the R-Q plane of Figure 14.7. The upper part of the surface corresponds to an outbreak level while the lower part corresponds to a subsistence level.

The modules to solve for the dynamics of the spruce budworm population are shown in Figures 14.8 to 14.10.

We drive changes in the model by setting X = TIME. The functions G(X) and F(X) generated by our STELLA model are shown in Figures 14.11 and 14.12.

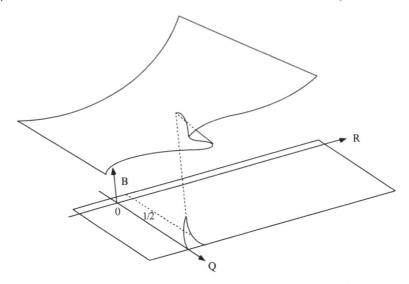

Fig. 14.7

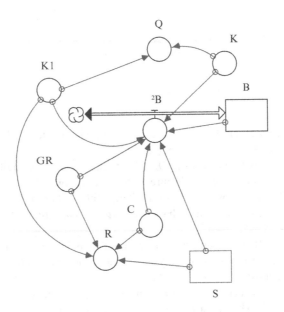

Fig. 14.8

For KS = 2.5, the corresponding values of X are 0.69, 2.0, and 7.32. These correspond to B = 0.173, 0.500, and 1.83, respectively (see equation (14.12)). These B values represent the steady states of B. However, only two of these extrema are

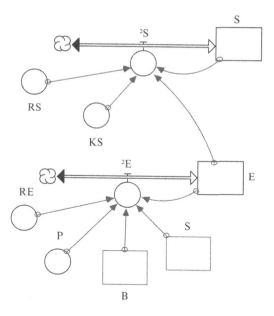

Fig. 14.9

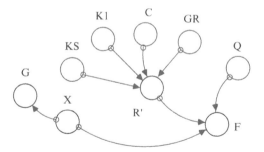

Fig. 14.10

stable: the middle value represents an unstable extremum. All initial values of B within a given range will lead to the same, single steady state for B, one of the two (see Figures 14.13 to 14.16. There are two such given ranges for initial Bs, given the way that the main model is set up—initial values of B greater and less than 1/2.

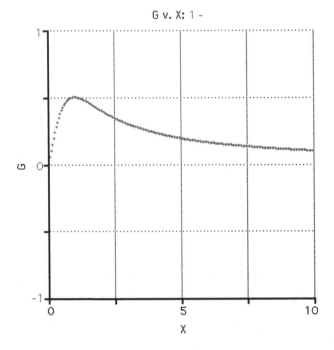

Fig. 14.11

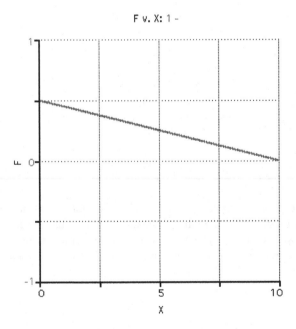

Fig. 14.12

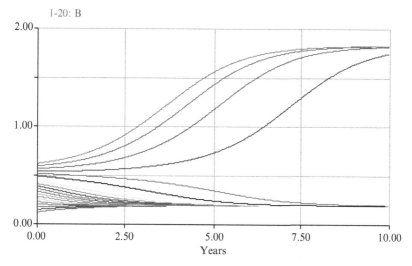

Fig. 14.13

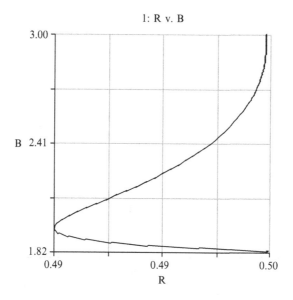

Fig. 14.14

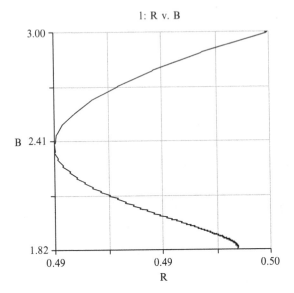

Fig. 14.15

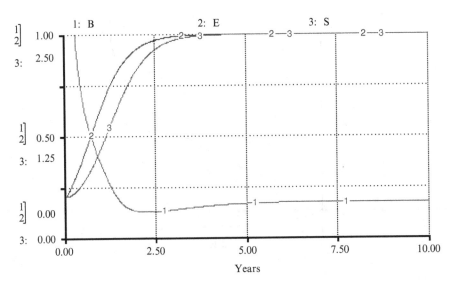

Fig. 14.16

14.3 Questions and Tasks

1. In the spruce-budworm model of Section 14.2, recognize that the choice of KS is
   a crucial one. So is the choice of Q. They determine whether one, two, or three

extrema solutions for B exist. Run a series of sensitivity tests on KS and Q to explore their impact on the system's dynamics.

2. (a) Change the initial conditions for the spruce-budworm dynamics model in a series of sensitivity runs. Explain the results.

   (b) Introduce an exogenous shock into your model that (at a random time) re-sets the budworm population to its initial condition. What are the impacts on the overall results? Explain your findings.

SPRUCE BUDWORM DYNAMICS

$B(t) = B(t - dt) + (\Delta B) * dt$
INIT $B = 3$ {Spruce Budworm per Unit Area}
INFLOWS:
$\Delta B = GR * B * (1 - B/K/S) - C * B^2/((K1 * S)^2 + B^2)$ {Spruce Budworm per Unit Area per Time Period}

$E(t) = E(t - dt) + (\Delta E) * dt$
INIT $E = .95$ {Percentage of Foliage Cover}
INFLOWS:
$\Delta E = RE*E*(1 - E) - P*B*E^2/S$ {Change in Percentage of Foliage Cover per Time Period}

$S(t) = S(t - dt) + (\Delta S) * dt$
INIT $S = 2.5$ {Habitat Density}
INFLOWS:
$\Delta S = RS * S * (1 - S/KS/E)$ {Habitat Density Change per Time Period}

$C = 1$
$F = R' * (1 - X/Q)$
$G = X/(1+X^2)$
$GR = 2$ {Spruce Budworm per Unit Area per Spruce Budworm per Unit Area per Time Period}
$K = 1$
$K1 = .1$
$KS = 2.5$
$P = .01$
$Q = K/K1$
$R = GR * K1 * S/C$
$R' = GR * K1 * KS/C$
$RE = 2$ {Change in Percentage of Foliage Cover per Percentage Foliage Cover Time Period}
$RS = 3$ {Habitat Density Change per Habitat Density Change per Time Period}
$X = TIME$

Chapter 15
Spatial Pestilence Dynamics

15.1 Diseased and Healthy Immigrating Insects

This chapter expands and refines some earlier models, which included life stages of insects, by specifically distinguishing two cohorts of a population infected with a disease. The two populations modeled here are insects that suffer from a disease that increases mortality for the infected nymphs and adults and also decreases their egg-laying rate. Unlike the previous chapters, we assume two populations of insects living in two fields. One of the fields has generally better living conditions than the other, although a current year's carrying capacities are randomly generated and there is some overlap in the ranges within which carrying capacities fluctuate.

The carrying capacities of the two fields have a direct effect on birth rates. The carrying capacities of the two fields are defined as

$$K1 = IF\,CARRY\_R1 > .666\,THEN\,2$$
$$ELSE\ IF\ CARRY\ R1 < .333\,THEN\,.5 \qquad (15.1)$$
$$ELSE\,1$$

and

$$K2 = IF\,CARRY\,R2 > .666\,THEN\,4$$
$$ELSE\ IF\ CARRY\ R2 < .333\,THEN\,1 \qquad (15.2)$$
$$ELSE\,2$$

respectively, with CARRY R1 and CARRY R2 as random numbers between 0 and 1. These random numbers are calculated in the following module with

$$R\,COUNT1 = IF\,MOD\,(TIME, 52) = 0\ THEN\ RANDOM(0,1)/DT\ ELSE\ 0 \qquad (15.3)$$

and

$$R\ COUNT2 = IF\,MOD\,(TIME, 52) = 0\ THEN\ RANDOM(0,1)/DT\ ELSE\ 0. \qquad (15.4)$$

B. Hannon and M. Ruth, *Dynamic Modeling of Diseases and Pests*,
Modeling Dynamic Systems,
© Springer Science + Business Media LLC 2009

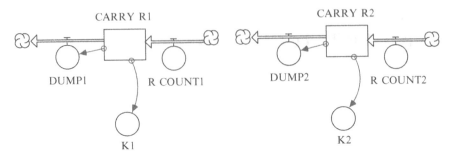

Fig. 15.1

where MOD is the built-in function that computes the remainder (modulo) of a division—in our case the division of TIME by 52 (Figure 15.1). We make use of the MOD function here to set up a recurring counter. Note that with some DT values, whose fractional representation does not have n^2 in the denominator, STELLA rounds the remainder in the MOD function; so the re-starting values of R COUNT1 and R COUNT2 for each new year are not exactly zero.

When overcrowding develops, healthy adult insects leave their home field and join the other population. Furthermore, it is assumed that 10% of healthy adults migrate under all circumstances. Changes in population sizes are no longer only dependent on births and on deaths but additionally on migration.

The model is composed of the following additional modules. The first captures the population dynamics of healthy insects in the first field (Figure 15.2).

The second module is set up to calculate the change in nymph and adult population in field 1 that are affected by the disease (Figure 15.3).

A virtually identical second set of these modules captures the dynamics of the populations in field 2. Parameters relevant to both healthy and diseased insects in both fields are calculated in the following modules. They include

- a calculation of the total number of adults in each fields, ALL ADULTS 1 and ALL ADULTS 2;
- the ratio of the total number of adults in each region to the carrying capacity of the respective region, FRXNL CAP1 and FRXNL CAP2;
- experimental maturation rates for healthy and diseased insects, F1 H, F1 D;
- model maturation rates U1 H, U1 D;
- experimental laying rates A2 H, A2 D;
- experimental daily adult survival fractions per stage, S2 H , S2 D; and
- adult mortality rates B1 H, B1 D, B2 H, B2 D.

The latter are calculated in the module of Figure 15.4, using the following exponential functions:

$$B1H = (1 - EXP(LOGN(S1H) * F1H * DT))/DT \qquad (15.5)$$

$$B1D = (1 - EXP(LOGN(S1D) * F1D * DT))/DT \qquad (15.6)$$

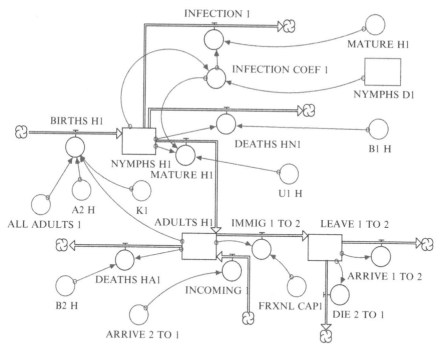

Fig. 15.2

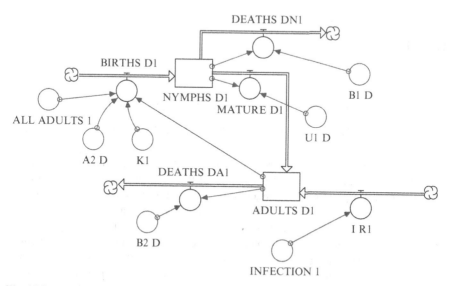

Fig. 15.3

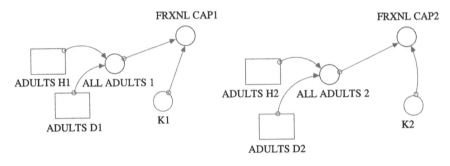

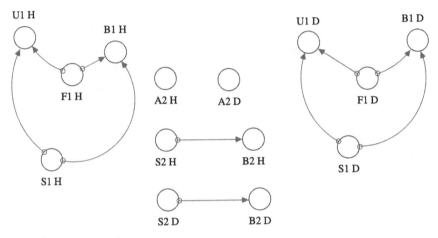

Fig. 15.4

$$B2H = (1 - EXP(LOGN(S2H)*DT/1))/DT \qquad (15.7)$$

$$B2D = (1 - EXP(LOGN(S2D)*DT/1))/DT \qquad (15.8)$$

Figure 15.5 shows the combined result of the population dynamics due to natural increases and deaths as well as migration. Over the long run, the population of each field is clearly responding to changes in the local carrying capacity. Numbers are slow to build up in both fields when the start is 0.1 healthy and diseased adults in each field. Surprisingly, neither field's total population hugs the carrying capacity very well in this graphed run, although that effect was evident in some other runs. Clearly factors other than carrying capacity are important. While we have required that 10% of healthy adults migrate, we are not seeing the diseased population expand to fill that gap.

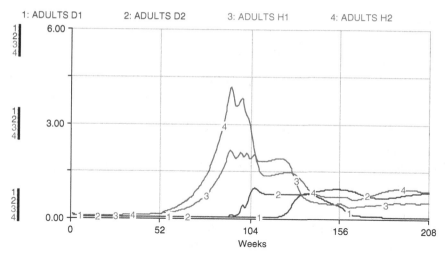

Fig. 15.5

15.1.1 Questions and Tasks

1. Note how ensuing runs of the model in this chapter—which use the same parameters and initial conditions—are significantly different. Why is this difference occurring?
2. In the model, the diseased insects are at a "disadvantage." What would you change (within the realm of the biologically likely) to favor the diseased population?
3. Is it possible that we can provide a more useful means for biological control by studying insect population dynamics from an ecological perspective? Can you implement such a control in the model?
4. Introduce additional factors, such as seasonality, into the population model.

DISEASED AND HEALTHY IMMIGRATING INSECTS

ADULTS_D1(t) = ADULTS_D1(t − dt) + (MATURE_D1 + I_R1 −
DEATHS_DA1) * dt
INIT ADULTS_D1 = .1 {Initial diseased adults.}

INFLOWS:
MATURE_D1 = U1_D * NYMPHS_D1 {Individuals per Time Period}
I_R1 = INFECTION_1 {Individuals per Time Period}
OUTFLOWS:
DEATHS_DA1 = ADULTS_D1 * B2_D {Individuals per Time Period}
ADULTS_D2(t) = ADULTS_D2(t − dt) + (MATURE_D2 + I_R2 −
DEATHS_DA2) * dt
INIT ADULTS_D2 = .1 {Initial diseased adults.}

INFLOWS:

MATURE_D2 = U1_D * NYMPHS_D2 {Individuals per Time Period}

I_R2 = INFECTION_2 {Individuals per Time Period}

OUTFLOWS:

DEATHS_DA2 = ADULTS_D2 * B2_D {Individuals per Time Period}

ADULTS_H1(t) = ADULTS_H1(t − dt) + (MATURE_H1 + INCOMING_1 − DEATHS_HA1 − IMMIG_1_TO_2) * dt

INIT ADULTS_H1 = .1 {Initial healthy adults}

INFLOWS:

MATURE_H1 = U1_H * NYMPHS_H1 * (1 − INFECTION_COEF1) {Individuals per Time Period}

INCOMING_1 = ARRIVE_2_TO_1 {Individuals per Time Period}

OUTFLOWS:

DEATHS_HA1 = ADULTS_H1 * B2_H {Individuals per Time Period}

IMMIG_1_TO_2 = IF ADULTS_H1 − (.1 * ADULTS_H1 + .9 * FRXNL_CAP1) > 0 THEN (.1 * ADULTS_H1 + .9 * FRXNL_CAP1) ELSE IF ADULTS_H1 > 0 THEN ADULTS_H1 ELSE 0 {Individuals per Time Period; Only healthy adults migrate. At least 10% of the healthy adults always migrate. Under the noted conditions the 10% healthy and an additional fraction of the healthy adults based empirically on the total number of adults also migrate. Note well the order of the nested IF statement; the first one is checked first and if the condition holds, the first statement is executed and the program goes no further. Otherwise all the adults flee. This same statement is also true of the adults in the other field.}

ADULTS_H2(t) = ADULTS_H2(t − dt) + (MATURE_H2 + INCOMING_2 − DEATHS_HA2 − IMMIG_2_TO_1) * dt

INIT ADULTS_H2 = .1 {Initial healthy adults}

INFLOWS:

MATURE_H2 = U1_H * NYMPHS_H2 * (1 − INFECTION_COEF2) {Individuals per Time Period}

INCOMING_2 = ARRIVE_1_TO_2 {Individuals per Time Period}

OUTFLOWS:

DEATHS_HA2 = ADULTS_H2 * B2_H {Individuals per Time Period}

IMMIG_2_TO_1 = IF ADULTS_H2 − (.1 * ADULTS_H2 + .9 * FRXNL_CAP2) > 0 THEN (.1 * ADULTS_H2 + .9 * FRXNL_CAP2) ELSE IF ADULTS_H2 > 0 THEN ADULTS_H2 ELSE 0 {Individuals per Time Period}

CARRY_R1(t) = CARRY_R1(t − dt) + (R_COUNT1 − DUMP1) * dt

INIT CARRY_R1 = 0

INFLOWS:

R_COUNT1 = IF MOD(time,52) = 0 THEN RANDOM(0,1)/DT ELSE 0

OUTFLOWS:

DUMP1 = IF MOD(time,52) = 0 THEN CARRY_R1/DT ELSE 0 {Insures a new number between 0–1 each integer time step.}

CARRY_R2(t) = CARRY_R2(t − dt) + (R_COUNT2 − DUMP2) * dt
INIT CARRY_R2 = 0 {See note in Carry_R.}

INFLOWS:
R_COUNT2 = IF MOD(time,52) = 0 THEN RANDOM(0,1)/DT ELSE 0
OUTFLOWS:
DUMP2 = IF MOD(time,52) = 0 THEN CARRY_R2/DT ELSE 0
LEAVE_1_TO_2(t) = LEAVE_1_TO_2(t − dt) + (IMMIG_1_TO_2 −
DIE_2_TO_1 − ARRIVE_1_TO_2) * dt
INIT LEAVE_1_TO_2 = 0

INFLOWS:
IMMIG_1_TO_2 = IF ADULTS_H1 − (.1 * ADULTS_H1 + .9 *
FRXNL_CAP1)
> 0 THEN (.1 * ADULTS_H1 + .9 * FRXNL_CAP1) ELSE IF ADULTS_H1 >
0 THEN ADULTS_H1 ELSE 0 {Individuals per Time Period; Only healthy
adults migrate. At least 10% of the healthy adults always migrate. Under the
noted conditions the 10% healthy and an additional fraction of the healthy adults
based empirically on the total number of adults also migrate. Note well the order
of the nested IF statement; the first one is checked first and if the condition holds,
the first statement is executed and the program goes no further. Otherwise all the
adults flee. This same statement is also true of the adults in the other field.}
OUTFLOWS:
DIE_2_TO_1 = .25 * LEAVE_1_TO_2 {Individuals per Time Period}
ARRIVE_1_TO_2 = .75 * LEAVE_1_TO_2 {Individuals per Time Period}
LEAVE_2_TO_1(t) = LEAVE_2_TO_1(t − dt) + (IMMIG_2_TO_1 − DIE_2TO1
− ARRIVE_2_TO_1) * dt
INIT LEAVE_2_TO_1 = 0

INFLOWS:
IMMIG_2_TO_1 = IF ADULTS_H2 − (.1 * ADULTS_H2 + .9 *
FRXNL_CAP2)
> 0 THEN (.1 * ADULTS_H2 + .9 * FRXNL_CAP2) ELSE IF ADULTS_H2 >
0 THEN ADULTS_H2 ELSE 0 {Individuals per Time Period}
OUTFLOWS:
DIE_2TO1 = .25 * LEAVE_2_TO_1 {Individuals per Time Period}
ARRIVE_2_TO_1 = .75 * LEAVE_2_TO_1 {Individuals per Time Period}
NYMPHS_D1(t) = NYMPHS_D1(t − dt) + (BIRTHS_D1 − DEATHS_DN1 −
MATURE_D1) * dt
INIT NYMPHS_D1 = 0 {Initial diseased eggs}

INFLOWS:
BIRTHS_D1 = IF (K1 − ALL_ADULTS_1) > 0 THEN A2_D * ADULTS_D1
ELSE 0 {Individuals per Time Period}
OUTFLOWS:
DEATHS_DN1 = B1_D * NYMPHS_D1 {Individuals per Time Period}

MATURE_D1 = U1_D * NYMPHS_D1 {Individuals per Time Period}
NYMPHS_D2(t) = NYMPHS_D2(t − dt) + (BIRTHS_D2 − DEATHS_DN2 −
MATURE_D2) * dt
INIT NYMPHS_D2 = 0 {Initial diseased eggs}

INFLOWS:
BIRTHS_D2 = IF (K2 − ALL_ADULTS_2) > 0 THEN A2_D * ADULTS_D2
ELSE 0 {Individuals per Time Period}
OUTFLOWS:
DEATHS_DN2 = B1_D * NYMPHS_D2 {Individuals per Time Period}
MATURE_D2 = U1_D * NYMPHS_D2 {Individuals per Time Period}
NYMPHS_H1(t) = NYMPHS_H1(t − dt) + (BIRTHS_H1 − DEATHS_HN1 −
MATURE_H1 − INFECTION_1) * dt
INIT NYMPHS_H1 = 0 {Initial Healthy eggs}

INFLOWS:
BIRTHS_H1 = IF (K1 − ALL_ADULTS_1) > 0 THEN A2_H * ADULTS_H1
ELSE 0 {Individuals per Time Period}
OUTFLOWS:
DEATHS_HN1 = B1_H * NYMPHS_H1 {Individuals per Time Period}
MATURE_H1 = U1_H * NYMPHS_H1 * (1 − INFECTION_COEF1)
{Individuals per Time Period}
INFECTION_1 = INFECTION_COEF1 * MATURE_H1 {Individuals per Time
Period}
NYMPHS_H2(t) = NYMPHS_H2(t − dt) + (BIRTHS_H2 − DEATHS_HN2 −
MATURE_H2 − INFECTION_2) * dt
INIT NYMPHS_H2 = 0 {Initial Healthy eggs}

INFLOWS:
BIRTHS_H2 = IF (K2 − ALL_ADULTS_2) > 0 THEN A2_H * ADULTS_H2
ELSE 0 {Individuals per Time Period}
OUTFLOWS:
DEATHS_HN2 = B1_H * NYMPHS_H2 {Individuals per Time Period}
MATURE_H2 = U1_H * NYMPHS_H2 * (1 − INFECTION_COEF2)
{Individuals per Time Period}
INFECTION_2 = INFECTION_COEF2 * MATURE_H2 {Individuals per Time
Period}
A2_D = .35 {Experimental laying rate. DISEASED EGGS PER ADULT PER
DAY.}
A2_H = .75 {Experimental laying rate. EGGS PER ADULT PER DAY.}
ALL_ADULTS_1 = ADULTS_H1 + ADULTS_D1
ALL_ADULTS_2 = ADULTS_H2 + ADULTS_D2
B1_D = (1 − EXP(LOGN(S1_D) * F1_D * DT))/DT {Egg mortality rate,
1/DAY. Instantaneous survival fraction + instantaneous mortality fraction = 1.}
B1_H = (1 − EXP(LOGN(S1_H) * F1_H * DT))/DT {Egg mortality rate,
1/DAY. Instantaneous survival fraction + instantaneous mortality fraction = 1.}

B2_D = (1 − EXP(LOGN(S2_D) * DT/1))/DT {Adult mortality rate, 1/day. One day = T = 1 = experimental period for which adult mortality is measured. Instantaneous survival fraction + instantaneous mortality fraction = 1.}

B2_H = (1 − EXP(LOGN(S2_H) * DT/1))/DT {Adult mortality rate, 1/day. One day = T = 1 = experimental period for which adult mortality is measured. Instantaneous survival fraction + instantaneous mortality fraction = 1.}

F1_D = .2 {Experimental maturation rate, 1/DAY, i.e., 20 eggs per 100 eggs mature each day, as noted in the experiment. In other words, a surviving egg matures on the average in five days under the experimental conditions.}

F1_H = .2 {Experimental maturation rate, 1/DAY, i.e., 20 eggs per 100 eggs mature each day, as noted in the experiment. In other words, a surviving egg matures on the average in five days under the experimental conditions.}

FRXNL_CAP1 = ALL_ADULTS_1/K1

DOCUMENT: This fraction is the degree to which both healthy and diseased adults have reached their carrying capacity.

FRXNL_CAP2 = ALL_ADULTS_2/K2

INFECTION_COEF1 = 1 − EXP(−.3 * NYMPHS_H1 * NYMPHS_D1) {Constructed function giving the desired 0 to 1 probability.}

INFECTION_COEF2 = 1 − EXP(−.3 * NYMPHS_H2 * NYMPHS_D2) {Constructed function giving the desired 0 to 1 probability.}

K1 = IF CARRY_R1 > .666
    THEN 2
  ELSE IF CARRY_R1 < .333
    THEN .5
  ELSE 1 {This is the carrying capacity of the area the insects are in}

K2 = IF CARRY_R2 > .666
    THEN 4
  ELSE IF CARRY_R2 < .333
    THEN 1
  ELSE 2 {This is the carrying capacity of the area the insects are in}

S1_D = .5 {Experimental diseased egg survival fraction, dimensionless, per stage. Stage = 1/F1, i.e., 30 eggs per 100 eggs survive each 1/F1 days, as noted in the experiment.}

S1_H = .7 {Experimental egg survival fraction, dimensionless, per stage. Stage = 1/F1, i.e., 70 eggs per 100 eggs survive each 1/F1 days, as noted in the experiment.}

S2_D = .65 {Experimental daily diseased adult survival fraction per stage, dimensionless.}

S2_H = .8 {Experimental daily adult survival fraction per stage, dimensionless.}

U1_D = F1_D * EXP(LOGN(S1_D) * F1_D * DT) {Model maturation rate for survivors, 1/DAY. Hatch rate = instantaneous survival fraction * maturation rate.}

U1_H = F1_H * EXP(LOGN(S1_H) * F1_H * DT) {Model maturation rate for survivors, 1/DAY. Hatch rate = instantaneous survival fraction * maturation rate.}

15.2 The Spatial Dynamic Spread of Rabies in Foxes[1]

15.2.1 Introduction

A spatially explicit computer model is developed to examine the dynamic spread of fox rabies across the state of Illinois and to evaluate possible disease control strategies. The ultimate concern is that the disease will spread from foxes to humans through the pet population. We are also concerned about the significant loss of an indigenous species.

Modeling the population dynamics of rabies in foxes requires comprehensive ecological and biological knowledge of the fox and pathogenesis of the rabies virus. Variables considered, including population densities, fox biology, home ranges, dispersal rates, contact rates, and incubation periods, can greatly affect the spread of disease. Accurate reporting of these variables is paramount for realistic construction of a spatial model. The spatial modeling technique utilized is a grid-based approach that combines the relevant geographic condition of the Illinois landscape (typically described in a georeferenced database system) with a nonlinear dynamic model of the phenomena of interest in each cell, interactively connected to the other appropriate cells (usually adjacent ones).

The resulting spatial model graphically links data obtained from previous models, fox biology, rabies information, and landscape parameters using various hierarchical scales and makes it possible to follow the emergent patterns. It also facilitates experimental stimulus/result data collection techniques. Results from the model indicate that the disease would likely spread among the native healthy fox population from east to west and would occur in epidemiological waves radiating from the point of introduction; becoming endemic across the state of Illinois in about 15 to 20 years. Findings also include the realization that while current hunting pressures can potentially extirpate the Illinois fox population, some level of hunting pressure could be used to control the disease.

Spatially explicit modeling of complex environmental problems is essential for developing realistic descriptions of past behavior and the possible impacts of

---

[1] Condensed from: A Dynamic Model of the Spatial Spread of an Infectious Disease: The Case of Fox Rabies in Illinois, with Brian Deal, Cheryl Farello, Mary Lancaster, Thomas Kompare, Bruce Hannon, Environmental Modeling and Assessment, 5:47–62, 2000.

alternative management policies[2]. Past ecosystem-scaled model development has been limited by the conceptual complexity of formulating, building, and calibrating intricate models. This has led to a general recognition of the need for collaborative modeling projects[3]. A graphically based, spatial modeling environment (SME) has been developed at the University of Maryland to address the conceptual complexity and collaborative barriers to spatio-temporal ecosystem model development. The modeling environment links icon-based graphical modeling environments (e.g. STELLA) with parallel supercomputers and a generic object database[4]. It allows users to create and share modular, reusable model components, and utilize advanced parallel computer architectures without having to invest unnecessary time in computer programming or learning new system.

The reader could run this complex model after learning the associated special software. However the process is somewhat complex and requires special hardware as well. Nevertheless, the model is included here in part as a demonstration of advanced modeling of diseases and pests that builds on the lessons from previous chapters in this book and to prompt exploration of more complex spatial dynamics that underlie many disease and pest issues.

15.2.2 Fox Rabies in Illinois

The epidemiology of fox rabies is intimately linked with fox behavior. Foxes produce their young in spring and juveniles migrate each fall and early winter. Adults will also migrate out of their home range if their population density is sufficiently high. This migratory behavior becomes the vehicle for widespread transmission of

[2] Risser, P.G., J.R. Karr, and R.T. Foreman. 1984. "Landscape Ecology: Directions and Approaches." Illinois Natural History Survey special publication; no. 2, Illinois Natural History Survey.

Costanza, R., F.H. Sklar, and M.L. White, 1990. BioScience 40 91–107.

Sklar, F.H. and R. Costanza. 1991. Quantitative Methods in Landscape Ecology, eds. M.G. Turner and R. Gardner, Springer-Verlag, New York, NY. pp. 239–288.

[3] Goodall, D.W. 1974. The Hierarchical Approach to Model Building, The First International Congress of Ecology, Wageningen, Netherlands, Center for Agricultural Publishing and Documentation.

Acock B. and J.F. Reynolds. 1990. Process Modeling of Forest Growth Responses to Environmental Stress, eds. R.K. Dixon, R.S. Meldahl, G.A. Ruark, and W.G. Warren, Timber Press, Portland, OR.

[4] Costanza, R. and T. Maxwell,. 1991. Ecological Modeling 58 159–183.

Maxwell, T. and R. Costanza. 1994. Toward Sustainable Development: Concepts, Methods, and Policy, vol. 58, eds. J. Van den Bergh and J. Van der Straaten, Island Press, Washington, D.C. pp. 111–138.

Maxwell, T. and R. Costanza. 1995. International Journal of Computer Simulation: Special Issue on Advanced Simulation Methodologies 5 247–262.

SME, http://kabir.cbl.umces.edu/SME3/index.html, International Institute for Ecological Economics, Center for Environmental Science, University of Maryland System (1999).

disease because the behavior of the infectious animal becomes erratic and combative, and the disease is then spread during contact with healthy foxes through biting. The incubation period for fox rabies varies from 14 to 90 days, ending in clinical illness. An animal may be infectious for up to a week before the onset of symptoms and remains infectious until death. Model parameters such as the effective biting rate and the actual length of the infected and infectious periods are difficult to determine in the field. We have used the best available data and determined an effective biting coefficient by trial and error comparisons of fox densities gained from the literature cited below.

A complete model of the fox and fox behavior might include a set of sex differentiated age cohorts. We found however, that the history of the disease and of fox behavior could be adequately represented by a simple four stock model of both healthy and sick juveniles and adults. The model includes both deterministic and stochastic components and can be adapted to any disease that possesses spatial dynamics by simply adjusting the input data. The results of our epidemic model indicate that the incidence of fox rabies can be decreased with an intervention strategy such as hunting. However, the results also indicate that the current fox hunting pressures, coupled with the introduction of the rabies disease, would lead to elimination of the fox in Illinois. Our results suggest that a reduced hunting pressure can leave a sustainable fox population in spite of the occasional introduction of the disease from surrounding areas. The disease can also be controlled by aerial deposition of baited vaccines over a large area. The model indicates the spatial dynamics of diseased foxes, and thus allows the most judicious and least expensive aerial deployment of the vaccine.

15.2.3 Previous Fox Rabies Models

Dynamic models of rabies in wildlife populations have been proposed by others[5]. These models have focused on the spatial spread of disease and potential impact of various control measures. But the addition of a spatial component to the disease dynamic is, in our opinion, a critical component. Spatial components can more easily explain variation in the rate of disease spread through a population[6], as well as provide a more holistic view of the dynamic interaction of animal, disease, and landscape. Since wildlife populations are not indolent and are typically in a perpetual state of flux, contact rates between diseased and healthy animals depend to some

[5] David, J.M., L. Andral, and M. Artois. 1982. Ecological Modeling 15 107–125.

  Bacon, P.J. 1985. Population Dynamics of Rabies in Wildlife, Academic Press, New York, NY.

  White, P.C.L., S. Harris, and G.C. Smith. 1995. Journal of Applied Ecology 32 693–706.

  Murray, J.D., E.A. Stanley, and D.L. Brown. 1986. Proceedings of the Royal Society of London B229 111–150.

  Gardner, G., A. Leslie, R.T. Gardner, and J. Cunningham 1990. Verlag der Zeitschrift fur Naturforschung 45c 1230–1240.

[6] Bacon, P.J. and D. MacDonald,. 1980. Nature 289 634–635.

extent on spatially derived information. David et al. (1982) proposed a simple model of vulpine rabies that included much of the same biological components we utilize in our model: reproduction, dispersal, and spatial distribution. The spatial component of the David model is not linked to habitat resources however, and the display mechanisms of the SME offer a much more explicit depiction of possible scenarios.

Vulpine rabies poses a serious problem in Europe due to increasingly large fox populations and its zoonotic potential, increasing the probability of human contact in heavily populated areas[7]. Fox densities in Bristol, England for example, range from 1.82 to 3.64 foxes per square kilometer over a home range size of 0.45 square kilometers[8]. In comparison, densities in the United States are lower: 0.15 foxes per square kilometer over a larger home range size of 9.6 square kilometers[9]. In the United States, rabies in the red fox, *Vulpes vulpes*, has reached epidemic levels in western Alaska and northern New York.

Previous linear models using data collected from European fox populations show a dramatic decrease in the number of foxes when rabies is introduced into a healthy population. These decreases reduce the population below an apparent disease threshold, and the disease is shown to die out[10]. These models typically demonstrate an inversely proportional relationship between infected and healthy foxes when rabies is first introduced into the population. As the disease becomes established, the number of infectious foxes increases and the susceptible fox population decreases[11]. Murray (1987) describes these density decreases as "breaks," where the population becomes too low for the disease to persist in the environment. Gardner et al. (1990) concluded that the disease could be eradicated from a fox population when fox numbers are reduced to a critical level below the carrying capacity. Most rabid epizootics do not drive fox populations to extinction however.

Several models demonstrated that both healthy and infected fox populations stabilized over a period of 20 to 30 years[12]. At this time, healthy populations reached levels that were half of the total carrying capacity and infected foxes were reduced below 10 percent of the total population[13]. Other models have concluded that rabies, a cyclical virus, can reemerge between 3.9 to 5 years after a period of quiescence[14].

We concluded that vulpine rabies could be viewed as a cyclical, nonlinear disease. When a susceptible population becomes infected, it decreases the healthy population but does not eliminate it. When the population rebuilds to a critical mass the disease is then able to reestablish, and the cycle begins again. In this way, vulpine

---

[7] Steck, F. and A. Wandeler. 1980. Epidemiologic Reviews 2 71–96.

[8] Trewhella, W.J., S. Harris, and F.E. McAllister. 1988. Journal of Applied Ecology 25 423–434.

[9] Storm, G.L., R.D. Andrews, R.L. Phillips, R.A. Bishop, D.B. Siniff, and J.R. Tester. 1976. Wildlife Monograph 49 1–81.

[10] White, et al. 1995; Murray, et al. 1986; Gardner, et al. 1990; Bacon and MacDonald. 1980; and: Murray, J.D. 1987. American Scientist 75 280–284.

[11] Murray, et al. 1986; Murray 1987; Gardner, et al. 1990.

[12] Gardner, et al. 1990; R.M. Anderson, H.C. Jackson, R.M. May, and A.M. Smith. 1981. Nature 289 765–771.

[13] Murray 1987; Anderson, et al. 1981.

[14] Murray et al. 1986; Gardner et al. 1990.

rabies is an epidemiological disease. This concept is important for the development of a spatial model that describes the spread of the disease over a landscape and for the evaluation of possible control measures.

15.2.4 The Rabies Virus

Canine rabies transmitted to humans has been reduced in the United States, although it is still a factor in over 75,000 human cases worldwide and is still considered a human health issue[15]. The disease, like many communicable diseases, appears to occur in cyclical waves. Its spread can best be understood through comprehensive study of the behavior of its mammalian hosts and the pathogenesis of the virus[16]. Typical host populations are heterogeneous in nature and field studies are difficult. There appears to be a hierarchy of susceptibility to rabies, with foxes, wolves, and coyotes being the most susceptible[17]. The fox adds to this complexity with shy and elusive behavior[18]. Although foxes do not typically interact with humans as frequently as other medium-sized mammals, they do come in contact with feral felines and canines. This contact increases the risk of stray cats and dogs contracting the rabies virus, and that risk places humans and domestic pets at risk.

The primary mode of rabies transmission is through the bite of an infected animal. To a lesser extent, scratching and licking can also transmit the disease. The virus replicates at the site of entry and once it reaches a sufficient titer the virus travels via the neural pathways to the brain[19]. The virus titer is defined as the smallest amount of virus per unit volume capable of producing infection[20]. The virus then travels from the central nervous system via peripheral nerves to the salivary glands, where it continues to multiply. Shedding of the virus in the saliva may occur before the appearance of clinical signs[21]. The incubation period, the time from inoculation to the appearance of clinical signs, can vary depending on the site of entry and its proximity to the central nervous system as well as the amount of virus entering the

---

[15] Fenner, F.J., E. Paul, J. Gibbs, F.A. Murphy, R. Rott, M.J. Studdert, and D.O. White, 1993. Veterinary Virology, Academic Press, New York, NY, 2nd ed.

Rupprecht, C.E., J.S. Smith, M. Fekadu, and J.E. Childs., Emerging Infectious Diseases 1. 1995. 107–114.

[16] Kaplan, C. , G.S. Turner, and D.A. Warrell. 1986. Rabies: The Facts, Oxford University Press, Oxford, England, ed. 2.

[17] MacDonald, D.W. 1980. Rabies and Wildlife: A Biologist's Perspective, Oxford Univ. Press, Oxford, UK.

Fields, B.N., D.M. Knipe, R.M. Chanock, M.S. Hirsch, J.L. Melnick, T.P. Monath, and B. Roizman. 1990. Virology, Raven Press, New York, NY, 2nd ed.

[18] MacDonald. 1980.

[19] Scherba, G. April 1998. Presentation to Ecological Modeling Group, Associate Professor of Veterinary Virology, University of Illinois.

[20] Fields et al. 1990; Scherba 1998; G.P. West, 1973. Rabies in Man and Animals, Arco Publishing Co, New York, NY.

[21] West 1973.

wound site[22]. Early clinical signs may be subtle and depend on where the virus is most concentrated in the central nervous system. Two clinical forms of rabies are known. The furious form affects the limbic system and, thus, the animal's behavior[23]. The dumb or paralytic form causes depression and lack of coordination[24]. Once clinical signs of rabies develop, death usually occurs in 7–10 days[25].

After the onset of clinical disease, foxes exhibit such overt behavioral changes as restlessness, pacing, and loss of appetite followed by either aggression or confusion, depending on the clinical form of the virus. The furious form of rabies will result in aggressive behavior, which encourages transmission of the disease. A fox with the paralytic form of rabies will become lethargic and confused and may only bite if provoked or approached by others. The final stages of either form of the disease are seizures and coma, followed by death. Normal foxes may "shy away" from rabid foxes thereby reducing their risk of infection[26].

15.2.5 Fox Biology

The red fox, *Vulpes vulpes*, is distributed throughout much of North America. Within the United States, the red fox has extended its range into forested areas where wolves and coyotes have been reduced or eliminated and where forests have been cleared[27].

The basic social unit of the red fox is typically a group of three or four breeding adults, and their juvenile offspring[28]. In cases where a territory includes several adults, one male and a variable number of closely related vixens are typically in a unit[29]. Average litter size is six pups and four pups generally survive until the time of dispersal[30].

Foxes are solitary, nocturnal foragers. They exploit the available food supply within a fairly well defined home range. Members of a group tend to follow each other from one resource patch to another and will eventually end up very close to the original point of departure toward the end of each night[31]. The vixens have individual ranges that overlap with each other and are encompassed by the home range

---

[22] Fields et al. 1990.

[23] West 1973.

[24] MacDonald 1980.

[25] Bacon, P. and D. MacDonald. 1980. New Scientist 28).

[26] Baer, G.M. 1975. The Natural History of Rabies: Volume I, Academic Press, New York, NY.

[27] Storm et al. 1976; T.G. Scott. 1955. Illinois Natural History Survey Division, Biological Notes. No. 35.

[28] Doncaster, C.P. and D.W. MacDonald 1997. Journal of Zoology 241 73–87.

[29] MacDonald 1980.

[30] Storm et al. 1976.

[31] Doncaster, C.P. and D.W. MacDonald. 1997. Journal of Zoology 241 73–87.

of the male fox, which essentially defines the territory of the group. Juveniles will exploit a limited number of resource patches close to their dens and within the home range of their parents, gradually expanding their ranges and separating themselves through the late summer until the time of dispersal[32]. Group ranges inevitably overlap to some degree, since an individual fox will normally cover less than half of its range in one night[33]. Encounters between foxes of different social groups in these overlapping areas will undoubtedly result in territorial conflicts when they occur at resource nodes.

Foxes that disperse from their home territories will normally travel in a relatively straight line until they find another territory that is available for occupancy[34]. If they are unable to find another territory within the dispersal season they become transients, forced to move continually. Hunting is the primary source of fox mortality, accounting for approximately 80% of deaths[35]. Hunting pressure is the primary mechanism for producing available territory during dispersal; that is when a majority of young foxes establish a home range. Hunting season in the state of Illinois extends from early November through the end of January.

Rabid foxes typically remain in their territories, but they do spend time resting at the peripheries, where they are more likely to come in contact with foxes from neighboring groups. Fox contact behavior however remains the most important unknown parameter in the spread of fox rabies[36]. Foxes exhibit different social behaviors, which may be density dependent. At lower densities, animals may be solitary or live in pairs. At higher densities, loose family groups occur and are generally comprised of one male, several females, and their offspring. Data from Sheldon[37] indicates that females determine the home range of a family group and that males are residents for only part of the year.

15.2.6 Model Design

A self-explanatory simplified flow chart of the spatial modeling procedure used in the development of the fox rabies model can be seen in Figure 15.6.

In this grid-based approach, each 6-mile square cell defined by the GIS maps is considered a typical fox home range. Each grid cell contains a highly nonlinear STELLA model that simulates the dynamic interaction and movement of foxes in one-month increments. The cellular model includes variables describing the propensity for fox immigration and emigration (based on fox population density) between adjacent cells. Each cellular model is automatically parameterized using the GIS

---

[32] Storm et al. 1976.

[33] Doncaster 1997.

[34] Storm et al. 1976.

[35] Storm et al. 1976.

[36] White, P.C.L. and S. Harris. 1994. Journal of Animal Ecology 63 315–327.

[37] Sheldon, W.G. 1950. Journal of Wildlife Management 14 33–42.

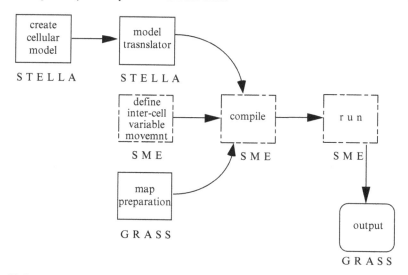

Fig. 15.6

maps for the area of concern. The GIS maps in our case referenced land-use conditions in Illinois that were used to develop fox carrying capacities in each of 1,610 cells that describe the state. Georeferenced maps were also used for the initial introduction of three diseased foxes along the state's eastern boundary (the disease appears to be spreading from east to west in the United States). The model collection is then run on a workstation computer for a model run-time of 25 years.

15.2.7 Cellular Model

A simplified STELLA model of fox population dynamics is shown in Figure 15.7. The four main variables measured—adult foxes, juvenile foxes, adult sick foxes, and juvenile sick foxes are represented as stock variables. The flow variables regulate the additions and subtractions to the stocks that take place at each time step (in this model, one month) and the rate variables help determine the amount of flow and changes in the flow variables. For the population dynamics model, flows and rates include: births of juvenile foxes, a death rate for each stock, emigration (out of) and immigration (into) each stock from adjoining cells, and a maturation of juvenile foxes into adulthood. A more detailed explanation of the model follows.

Details of the fox birth rates, maturation rates, infection rates, mortality rates, and migration rates are given in Deal et al. (2000)[38].

---

[38] Deal, B., C. Farello, M. Lancaster, T. Kompare and B. Hannon. 2000. A Dynamic Model of the Spatial Spread of an Infectious Disease: The Case of Fox Rabies in Illinois, Environmental Modeling and Assessment, Vol. 5, pp. 47–62.

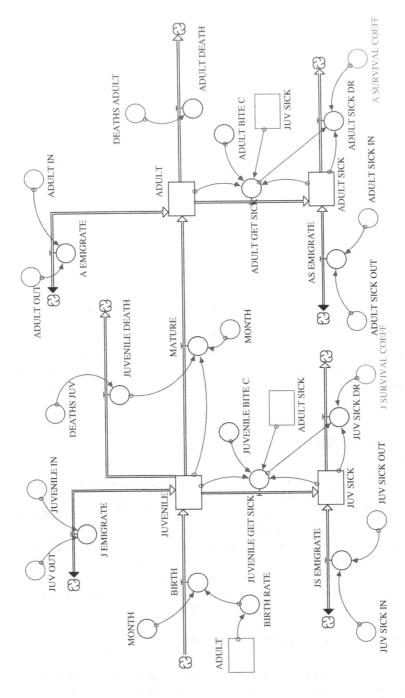

Fig. 15.7

15.2.8 Model Assumptions

As with all complex models, a series of assumptions is required. Here are ours:

1. There is no significant difference in the behavior of males and females regarding interaction outside the family unit and dispersion. The behavior of foxes is not significantly different when foxes outside the family unit are encountered.
2. Large rivers are not barriers to migration; it is assumed that foxes will use bridges, swim, or walk across the frozen surface during the winter.
3. Rabies is always fatal; every fox that contracts rabies will die[39].
4. All rabid foxes are equal sources of infection; all foxes during the rabid phase will behave similarly.
5. Rabies incubation and infectious periods are the same length of time; there is no significant difference in the time periods of incubation and being infectious.
6. All surviving juvenile foxes mature to adulthood at 11 months.
7. There is no significant difference in the amount of resources each fox uses, regardless of age.
8. The fox population has a definable maximum. We used the maximum densities of foxes found in Great Britain[40].
9. The model starts with 3 rabid foxes along the Illinois/Indiana border.

15.2.9 Georeferencing the Modeling Process

Red foxes live in a variety of habitats[41]. According to the Illinois Natural History Survey, foxes avoid forested areas and interior urban areas[42]. Foxes use forested areas for migration but forests are commonly avoided due to coyote predation. In general, red foxes are found in open croplands, grasslands, or pasture, using sloped areas for den sites. Urban edge areas and farmsteads are important habitats for the red fox due to the abundance of prey and forage in these areas. Gosselink estimated the average east-central Illinois red fox density at three adult foxes per 10 square miles; a breeding pair, plus one nonbreeding (juvenile) fox[43]. An average litter size is estimated at six pups, so approximately nine foxes are estimated to occupy every 10 square miles in Illinois.

This estimate of the average healthy carrying capacity in the state helped to create a habitat-weighted fox carrying capacity map for the state of Illinois. This was done using data collected by the Illinois Natural History Survey in the Land Cover of

---

[39] Bacon and MacDonald 1980.

[40] Anderson et al. 1981.

[41] Storm et al. 1976.

[42] Gosselink, T. March 1988. Personal Communication, Illinois Natural History Survey, Champaign, IL.

[43] Gosselink 1998.

Illinois GIS based mapping projects[44]. Urban edge cells were assigned a fox habitat value of 5 (urban edge is defined as any areas within 0.5 kilometers of an urban–non-urban interface) and all interior urban cells were assigned a value of 0. Forest cells were assessed a value of 1 and remaining cells, including wetlands, croplands, and grass/pasture lands were assigned a range of values from 2 to 3 depending on slope.

The Land Cover of Illinois map data was aggregated into 6- by 6-mile cells, statewide. In this new carrying capacity map, the minimum healthy carrying capacity cell value is 21.50 foxes per cell and the maximum healthy carrying capacity cell value is 70.84 foxes per cell. This map was exported into a GRASS raster based format and imported into the Spatial Modeling Environment (SME) for use with the STELLA fox model.

15.2.10 Spatial Characteristics

Adult foxes do not always remain in the territory they chose during juvenile dispersal. Adults are known to disperse only in the fall and winter, as juveniles do, so it is possible for incoming juveniles to displace some adult foxes. To simulate these characteristics and to simplify intercellular movement, the spatial model randomly assigns the direction of travel and limits travel to the four main compass directions. A fox that moves into a habitat cell that is at the carrying capacity (as derived from the map) forces the movement of a fox out of the same cell.

Emigration is a function of the spatial resource limits of each cell. In this model, the carrying capacity of a cell is a fixed amount and has been determined by land use cover characteristics. The relative population of foxes in each cell fluctuates with births, deaths, and immigration/emigration. Foxes emigrate when the calculated population of foxes in the cell exceeds the carrying capacity of that cell. This simulates the relationship between fox populations and resource availability. When the relative fox population of an area exceeds the resources available, there is pressure for a part of the population to move.

Once the total number and type of emigrants for each stock is determined, directional preferences must be calculated. Foxes sometimes travel great distances to find suitable and available habitat. It appears however, that in most instances foxes choose home ranges based on availability and not attractiveness. For this reason, a directional preference was assigned randomly for each group of emigrants. These random assignments occur only when certain landscape and time considerations are met. The directional assignment ratios are then applied to the stocks to determine the number and direction of foxes moving out of each cell at each time step.

Immigration, driven primarily by the emigration function, is the number of foxes in each of the four primary stocks that have been added to the cell at each computational time step. If the incoming foxes, plus foxes in the cell, exceed the carrying

---

[44] Luman, D. , M. Joselyn, and L. Suloway. 1996. "Illinois Scientific Survey Joint Report #3" Illinois Natural History Survey.

capacity of the cell then the emigration function is re-activated and the movement process begins anew. If the carrying capacity has not been exceeded then there is no pressure for any foxes to leave the base cell and none will emigrate.

15.2.11 Model Constraints

The following are the key constraints imposed on our model:

A. There is no interaction with the external geography of the model; foxes may not come or go beyond the model's defined boundaries.
B. Foxes may only migrate to another cell in the four cardinal directions.
C. The rabies disease may only be transmitted to another cell by a migrating fox. This implies that home ranges may only exist wholly within a cell, and may not overlap cell boundaries. There are no fractional home ranges within a cell.
D. There is no edge interaction between cells other than migration. Unlike, adjacent home ranges within cells, adjacent home ranges that happen to be in two different cells do not interact, unless migration occurs.

15.2.12 Model Results

The spatial dynamics of the spread of the fox rabies disease in Illinois provides an interesting picture. Figure 15.8 displays four maps produced from the full 25-year run of the model. The maps show the spread of the disease among the originally healthy, un-hunted fox population.

Since the disease has yet to hit the state, the only model calibration available was the historic rate of spread of the disease, about 24 miles per year[45]. The critical

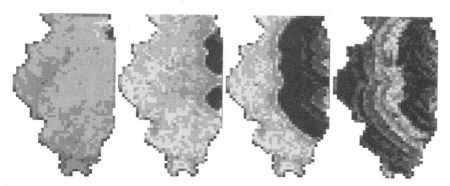

Fig. 15.8

---

[45] Gosselink (1998).

unknown parameter, the "Bite Coefficient"—the parameter that multiplies the product of the sick and healthy population,—was adjusted to give this rate of spread.

Moving from left to right, the initial map depicts the state of the system at present—that is, the initial number of healthy foxes across the state of Illinois—black areas are indicative of cells with low fox populations. The initial map also gives the reader an indication of the scale of the typical fox home range (each pixel in the picture represents the 6-square-mile cell size in the model) and the complexity of the computational problems involved.

The second map shows the incipient waves of disease that are created by the introduction of just three rabid foxes in the first month of the model run (these three rabid foxes are introduced into the model only once). Easily visible at the eastern border of the state, the dark areas in this map are areas of minimal fox survival. The map displays the dramatic impact that rabies can have on a healthy population.

After 2.6 years, the advance of the first wave of disease is clearly recognized. In the third map, the disease has spread halfway across the state, and is calculated to be advancing at about 24 miles per year. Also visible at this point is the reintroduction of the disease at its origin. This wave phenomena has also been noted in the construction of other dynamic epidemic disease models[46].

The fourth panel (7 years) describes a mature disease, and the wave of disease is in its epizootic stage. The traveling wave is continuously repeated, with each successive wave peak of the disease more spatially diffused in the east-west direction of the advance. This traveling epizootic wave can be seen as the slow cycle on the annual cycle in Figure 15.9. The successive waves develop because the disease does not eradicate the entire population and some of the disease remains viable in the surviving foxes. The disease is unable to move at this point, however, because the population of available foxes is not large enough to encourage much migration. Once the population builds to critical mass however, migratory behavior resumes and the disease begins to spread again. Perhaps augmented by the occasional healthy immigrant from Indiana, they re-grow to a substantial healthy population but again are infected by the strays heading east from the disease front. The epidemic builds more slowly in each ensuing repetition, since it begins with a smaller healthy population each time. Finally, although difficult to display in a static format, the disease becomes endemic—without epidemic pulse. In this stage, the number of diseased foxes in the state is nearly constant (5,586).

Mapped images are extremely powerful for displaying the spatial interactions and dynamic movement of the rabies disease. Although difficult to represent in static format, the animations of these images provide a strong case for the use of spatial simulation modeling for numerous applications. The mapped images will also become important in future work regarding the most efficient disease control strategies. A more quantified approach for displaying the results of our calibrated model runs follows.

---

[46] Hannon and Ruth 1997.

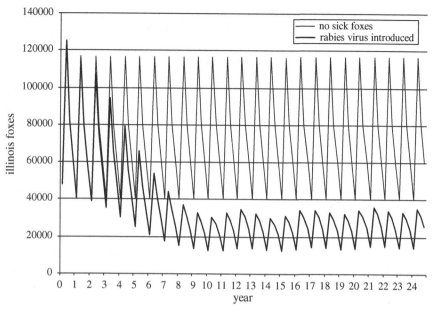

Fig. 15.9

15.2.13 Rabies Pressure

Although hunting pressure now exists on an apparently declining Illinois fox population, we thought the presentation of our results would be clearer if we started the
model with the fox population at or near the 1970 estimated mean level of about
88,000. We lowered the initial population in the model, but the long-term results of
subsequent runs were unaffected by this kind of change in the initial conditions. For
the first part of our model runs, we choose to suspend the effects of hunting to allow
us to gauge the impact of the disease alone.

Healthy fox populations were set initially at 80% of the carrying capacity of
the cell. The model results show that the healthy, un-hunted population settles to
a reasonable fluctuation between 40,000 and 118,000, with a mean level of 79,000
(Figure 15.4). The total population cycle shown in this figure is caused by the birth
of a large number of juveniles every March and the automatic redefinition of the
surviving juveniles to adults in the ensuing January.

15.2.14 The Effects of Disease Alone

Our next step was to introduce the disease to a healthy, cyclically stable, un-hunted
fox population. Since rabies is noted to be spreading westward from Pennsylvania,
we choose to introduce three rabid adult foxes in the first month of the 25-year run.

We choose to separate the effects of hunting from those of the disease. The results are shown in Figure 15.9. Note that the fox population declines to an average steady cycle with a mean of about 22,000 foxes, this is a 72 percent reduction from initial population levels. From this model run, our estimate of the effect of rabies is that the population will suffer a severe setback (a reduction to one-quarter of the healthy mean) but the fox does not disappear from the landscape. To ecologists, this is the good news: the fox is not eliminated by the disease. To public health officials, the news is not so good: diseased foxes remain in the state, albeit at much reduced levels; therefore, some threat of spread of the disease into the pet population remains.

15.2.15 Hunting Pressure

We then introduced the full effect of the estimated current level of hunting. The results indicate that the healthy fox population disappears over a 25-year period at current rates. Our hunting rates are constant from year to year however, an assumption which cannot be totally accurate. Some of the fox hunting is for pelts, and that portion (unknown) of the hunting pressure would fluctuate with the price of pelts and the availability of foxes.

The Illinois fox population, subjected to current hunting pressures, declines to essentially zero over the ensuing 25 years with the concurrent introduction of the disease. We made several more runs with the hunting pressure reduced to one-half and then one-quarter of the full-scale current hunting pressure. As one reduces the hunting pressure, the fox population approaches the no-hunting, cyclic result discussed earlier, , with most of the recovery made when the hunting pressure was halved. It was possible to show that a small amount of hunting will reduce the number of rabid foxes in the endemic stage of the disease.

15.2.16 Controlling the Disease

Control of wildlife disease is often expensive. Evaluation of control measures in dynamic modeling facilitates decision making and gives policy makers additional information and insight into the effects of the control measures tested. Use of dynamic models in the study of wildlife diseases also identifies areas where information is lacking.

Control attempts without clear knowledge of the spatial qualities of the studied population or disease can produce less than optimal results[47]. Bögel et al. suggested that methods used in advance of a rabies epizootic would not be effective in controlling the spread of the disease once the epizootic has begun[48]. Additional work

---

[47] Cowan, I.M. 1949. Journal of Mammalogy 30 396–398.

[48] Bögel, K., H. Moegle, W. Krocza, and L. Andral. 1981. Bulletin of the World Health Organization 59 269–279.

by Bögel, et al. resulted in a proposed method to evaluate populations and rabies control in the wild.

Anderson, et al. evaluated culling, vaccination, and culling combined with vaccination as control measures for vulpine rabies. Their model revealed that population dynamics, most notably reproduction, limited the effectiveness of culling alone[49]. Vaccination of foxes is a commonly chosen control method, but again, has had limited success, particularly in areas of high animal density and good habitat[50]. It is estimated that nearly 100% vaccination is required in areas of good habitat and greater than 15 foxes per square kilometer[51].

So vaccine programs (the current preferred method of control) are expensive and potentially ineffective. The cost-effectiveness of an airdrop, vaccinated bait program could be increased dramatically by more precise knowledge of the location and rate of spread of the disease. A modeling process such as this one should be of great help, assuming proper monitoring of the current status of the disease. The model presented can describe a reasonable estimate of the location and rate of spread of the disease front. The model would need to include some additional temporal variability, such as lags in reporting the appearance of rabid foxes, in the initiation of the vaccine drop, and between vaccine drop and discovery by the foxes, as well as the effective life of the bait. The model could be expanded to become effective in this arena if appropriate data were found.

FOX RABIES MODEL – BASIC MODEL COMPONENT

Adult Density Dependent Mortality

DEATHS_ADULT = IF CELL_POP > MIGRATORY_CC THEN
DR_DD_ADULT ELSE DR_NAT * ADULT + DR_HUNT * ADULT
DR_DD_ADULT = IF CELL_POP > 0 THEN DDD * ADULT/CELL_POP +
(DR_NAT + DR_HUNT) * (CELL_POP − DDD * DT) * ADULT/CELL_POP
ELSE 0

Fox Population Dynamics

ADULT(t) = ADULT(t − dt) + (MATURING + A_EMIGRATING −
ADULT_DYING − ADULT_SICKENING) * dt

INIT ADULT = (MIGRATORY_CC * .8)

INFLOWS:

MATURING = If (MONTH =1) AND (JUVENILE > JUVENILE_DEATH)
then (JUVENILE − JUVENILE_DEATH) else 0

A_EMIGRATING = (ADULT_IN − ADULT_OUT)

---

[49] Anderson et al. 1981.

[50] Baer, G.M., M.K. Abelseth, and J.G. Debbie. 1971. American Journal of Epidemiology 93 487–490. Black, J.G. and K.F. Lawson. 1973. Canadian Veterinary Journal 14 206–211.

[51] Anderson et al. (1981).

OUTFLOWS:
ADULT_DYING = (DEATHS_ADULT)
ADULT_SICKENING = (ADULT * ADULT_SICK * ADULT_BITE_C) +
(JUV_SICK * ADULT * ADULT_BITE_C)
ADULT_SICK(t) = ADULT_SICK(t − dt) + (ADULT_SICKENING +
AS_EMIGRATING − ADULT_SICK_DYING) * dt
INIT ADULT_SICK = MAP_INFECTED

INFLOWS:

ADULT_SICKENING = (ADULT * ADULT_SICK * ADULT_BITE_C) +
(JUV_SICK * ADULT * ADULT_BITE_C)
AS_EMIGRATING = (ADULT_SICK_IN − ADULT_SICK_OUT)
OUTFLOWS:
ADULT_SICK_DYING = ((ADULT_SICK * A_SURVIVAL_COEFF) −
ADULT_SICKENING)
JUVENILE(t) = JUVENILE(t − dt) + (BIRTHING + J_EMIGRATING −
JUV_SICKENING − JUVENILE_DEATH − MATURING) * dt
INIT JUVENILE = 0

INFLOWS:

BIRTHING = if MONTH = 3 then BIRTH_RATE else 0
J_EMIGRATING = (JUVENILE_IN − JUV_OUT)
OUTFLOWS:
JUV_SICKENING = (JUVENILE * JUV_SICK * JUVENILE_BITE_C) +
(ADULT_SICK * JUVENILE * JUVENILE_BITE_C)
JUVENILE_DEATH = (DEATHS_JUV)
MATURING = If (MONTH =1) AND (JUVENILE > JUVENILE_DEATH)
then (JUVENILE − JUVENILE_DEATH) else 0

JUV_SICK(t) = JUV_SICK(t − dt) + (JUV_SICKENING + JS_EMIGRATING
− JUV_SICK_DYING) * dt
INIT JUV_SICK = 0

INFLOWS:

JUV_SICKENING = (JUVENILE * JUV_SICK * JUVENILE_BITE_C) +
(ADULT_SICK * JUVENILE * JUVENILE_BITE_C)
JS_EMIGRATING = (JUV_SICK_IN − JUV_SICK_OUT)
OUTFLOWS:
JUV_SICK_DYING = (JUV_SICK * J_SURVIVAL_COEFF) −
JUV_SICKENING
ADULT_BITE_C = .015
A_SURVIVAL_COEFF = .33
BIRTH_RATE = ((0.95 * ADULT/2) * NORMAL(6.5, 0.5))
JUVENILE_BITE_C = .015
J_SURVIVAL_COEFF = .33

MONTH = MOD(time, 12) +1

Incoming Populations

ADULT_IN = A_GO_E@W + A_GO_N@S + A_GO_S@N + A_GO_W@E
ADULT_SICK_IN = AS_GO_E@W + AS_GO_N@S + AS_GO_S@N + AS_GO_W@E
AS_GO_E@W = 0
AS_GO_N@S = 0
AS_GO_S@N = 0
AS_GO_W@E = 0
A_GO_E@W = 0
A_GO_N@S = 0
A_GO_S@N = 0
A_GO_W@E = 0
JS_GO_E@W = 0
JS_GO_N@S = 0
JS_GO_S@N = 0
JS_GO_W@E = 0
JUVENILE_IN = J_GO_E@W + J_GO_N@S + J_GO_S@N + J_GO_W@E
JUV_SICK_IN = JS_GO_E@W + JS_GO_N@S + JS_GO_S@N + JS_GO_W@E
J_GO_E@W = 0
J_GO_N@S = 0
J_GO_S@N = 0
J_GO_W@E = 0

Juvenile Density Dependent Mortality

DDD = IF GAP_POP >0 THEN CELL_POP/MAX_DENSITY_CC *
GAP_POP/3 ELSE 0
DEATHS_JUV = IF (CELL_POP > MIGRATORY_CC) THEN (DR_DD_JUV)
ELSE (DR_NAT * JUVENILE) + (DR_HUNT * JUVENILE)
DR_DD_JUV = IF CELL_POP > 0 THEN DDD * JUVENILE/CELL_POP +
(DR_NAT + DR_HUNT) * (CELL_POP − DDD * DT) *
JUVENILE/CELL_POP ELSE 0

Natural Mortality

DR_HUNT = DR_HUNT_ADD * HUNT
DR_MULT = 1
DR_NAT = DR_NO_HUNT * DR_MULT
HUNT = 1
DR_HUNT_ADD = GRAPH(MONTH)
(1.00, 0.2), (2.00, 0.085), (3.00, 0.014), (4.00, 0.004), (5.00, 0.004), (6.00, 0.003), (7.00, 0.001), (8.00, 0.002), (9.00, 0.005), (10.0, 0.029), (11.0, 0.052), (12.0, 0.318)
DR_NO_HUNT = GRAPH(MONTH)

(1.00, 0.022), (2.00, 0.015), (3.00, 0.014), (4.00, 0.039), (5.00, 0.025), (6.00, 0.028), (7.00, 0.015), (8.00, 0.014), (9.00, 0.028), (10.0, 0.114), (11.0, 0.209), (12.0, 0.136)

Outgoing Direction Assignment

AM_EAST = IF AM_TOTAL > 0 THEN ARE/AM_TOTAL ELSE 0
AM_NORTH = IF AM_TOTAL > 0 THEN ARN/AM_TOTAL ELSE 0
AM_SOUTH = IF AM_TOTAL > 0 THEN ARS/AM_TOTAL ELSE 0
AM_TOTAL = ARE + ARN + ARS + ARW
AM_WEST = IF AM_TOTAL > 0 THEN ARW/AM_TOTAL ELSE 0
ARE = RANDOM(0,1)
ARN = RANDOM(0,1)
ARS = RANDOM(0,1)
ARW = RANDOM(0,1)
ASM_EAST = IF ASM_TOTAL > 0 THEN ASRE/ASM_TOTAL ELSE 0
ASM_NORTH = IF ASM_TOTAL > 0 THEN ASRN/ASM_TOTAL ELSE 0
ASM_SOUTH = IF ASM_TOTAL > 0 THEN ASRS/ASM_TOTAL ELSE 0
ASM_TOTAL = ASRE + ASRN + ASRS + ASRW
ASM_WEST = IF ASM_TOTAL > 0 THEN ASRW/ASM_TOTAL ELSE 0
ASRE = RANDOM(0,1)
ASRN = RANDOM(0,1)
ASRS = RANDOM(0,1)
ASRW = RANDOM(0,1)
JM_EAST = IF JM_TOTAL > 0 THEN JRE/JM_TOTAL ELSE 0
JM_NORTH = IF JM_TOTAL > 0 THEN JRN/JM_TOTAL ELSE 0
JM_SOUTH = IF JM_TOTAL > 0 THEN JRS/JM_TOTAL ELSE 0
JM_TOTAL = JRE + JRN + JRS + JRW
JM_WEST = IF JM_TOTAL > 0 THEN JRW/JM_TOTAL ELSE 0
JRE = RANDOM(0,1)
JRN = RANDOM(0,1)
JRS = RANDOM(0,1)
JRW = RANDOM(0,1)
JSM_EAST = IF JSM_TOTAL > 0 THEN JSRE/JSM_TOTAL ELSE 0
JSM_NORTH = IF JSM_TOTAL > 0 THEN JSRN/JSM_TOTAL ELSE 0
JSM_SOUTH = IF JSM_TOTAL > 0 THEN JSRS/JSM_TOTAL ELSE 0
JSM_TOTAL = JSRE + JSRN + JSRS + JSRW
JSM_WEST = IF JSM_TOTAL > 0 THEN JSRW/JSM_TOTAL ELSE 0
JSRE = RANDOM(0,1)
JSRN = RANDOM(0,1)
JSRS = RANDOM(0,1)
JSRW = RANDOM(0,1)

Outmigration/Population Totals

ADULT_PCT = IF CELL_POP >0 THEN ADULT/CELL_POP ELSE 0
ADULT_SICK_PCT = If CELL_POP > 0 then ADULT_SICK/CELL_POP else 0

CELL_POP = ADULT + ADULT_SICK + JUVENILE + JUV_SICK
EM_ADULT = (ADULT_PCT * TOTAL_MIGRANTS)
EM_ADULT_SICK = (ADULT_SICK_PCT * TOTAL_MIGRANTS)
EM_JUV = (JUV_PCT * TOTAL_MIGRANTS)
EM_JUV_SICK = (JUV_SICK_PCT * TOTAL_MIGRANTS)
JUV_PCT = IF CELL_POP >0 THEN JUVENILE/CELL_POP ELSE 0
JUV_SICK_PCT = if CELL_POP>0 then JUV_SICK/CELL_POP else 0
TOTAL_MIGRANTS = IF (CELL_POP − MIGRATORY_CC) <= 0 THEN 0
ELSE CELL_POP − MIGRATORY_CC

Quantifying the Outgoing Rates

ADULT_OUT = IF (MONTH >=9) AND (MONTH <= 11) THEN
EM_ADULT ELSE 0
ADULT_SICK_OUT = {If ADULT_SICK < 3 and ADULT_SICK >1 then 0
else} (ADULT_SICK/3)
AS_GO_E = (ADULT_SICK_OUT * ASM_WEST)
AS_GO_N = (ADULT_SICK_OUT * ASM_SOUTH)
AS_GO_S = (ADULT_SICK_OUT * ASM_NORTH)
AS_GO_W = (ADULT_SICK_OUT * ASM_EAST)
A_GO_E = (ADULT_OUT * AM_WEST)
A_GO_N = (ADULT_OUT * AM_SOUTH)
A_GO_S = (ADULT_OUT * AM_NORTH)
A_GO_W = (ADULT_OUT * AM_EAST)
JS_GO_E = (JSM_WEST * JUV_SICK_OUT)
JS_GO_N = (JSM_SOUTH * JUV_SICK_OUT)
JS_GO_S = (JSM_NORTH * JUV_SICK_OUT)
JS_GO_W = (JSM_EAST * JUV_SICK_OUT)
JUV_OUT = IF (MONTH >=9) AND (MONTH <= 11) THEN EM_JUV
ELSE 0
JUV_SICK_OUT = IF (MONTH >=9) AND (MONTH <= 11) THEN
EM_JUV_SICK ELSE 0
J_GO_E = (JM_WEST * JUV_OUT)
J_GO_N = (JM_SOUTH * JUV_OUT)
J_GO_S = (JM_NORTH * JUV_OUT)
J_GO_W = (JM_EAST * JUV_OUT)

Spatial Resource Limits

GAP_POP = CELL_POP − MIGRATORY_CC
MAP_BOUNDARY = 0
MAP_CARRYING_CAPACITY = 0
MAP_INFECTED = 0
MAP_MULTIPLIER = 2.234
MAX_DENSITY_CC = MIGRATORY_CC * 2.3
MIGRATORY_CC = MAP_CARRYING_CAPACITY * MAP_MULTIPLIER

Part III
Conclusions

Chapter 16
Conclusion

Being able to better understand how diseases and pests are propagated and how they may be contained requires knowledge of the many factors that influence their success or failure. The combination of chance events, time delays, and nonlinearities in complex social and environmental settings make predicting spread of diseases and pests a daunting task. Yet, some method for sorting through the myriad of factors and influences and for anticipating future dynamics is required to meet dual goals of improving the state of human welfare and maintaining healthy ecosystems.

In this book we provided one particular, and very powerful, way of looking at the world. We concentrated on the forces that underlie different dynamic systems. Others, with different educational or cultural backgrounds, may choose a different approach and may develop different models. The potential diversity of perspectives and approaches in modeling is a challenge for all of us and should be perceived as an opportunity to engage in cross-disciplinary and cross-cultural dialogue about the world in which we live.

STELLA, through its use of graphics, is an excellent tool to organize and communicate model assumptions, structure, and results among individuals with different backgrounds. You will soon find that your models become increasingly detailed. Frequently, model efforts become large-scale multidisciplinary endeavors. STELLA is sufficiently versatile to enable development of complex, large-scale dynamic models. Such models can include a variety of features that are typically not dealt with by an individual modeler. Through easy incorporation of new modules into existing dynamic models and flexibility in adjusting models to specific real-world problems, STELLA fosters dialog and collaboration among modelers. It is a superb organizing and knowledge-capturing device for model building in an interdisciplinary arena. Individuals can easily integrate their knowledge into a STELLA model without "losing sight" of, or influence on, their particular part of the model.

Even though the models developed in this book were guided toward an explanation of real-world phenomena, empirical applications are not the focus of this book. Nevertheless, once we developed sufficiently elaborate models, we made intensive use of real-world data. These and other models illustrate the applicability to real-world data and, in general, the power of the dynamic modeling approach chosen

B. Hannon and M. Ruth, *Dynamic Modeling of Diseases and Pests*,
Modeling Dynamic Systems,
© Springer Science+Business Media LLC 2009

in this book. We strongly encourage you to take up and refine some of the models presented in this book to further accommodate data from real systems.

We selected a variety of different systems, spanning the disciplines of physics, genetics, biology, ecology, economics, and engineering, to illustrate the power of dynamic modeling and the multitude of possible applications. Other dynamic modeling books geared specifically toward some of these and other disciplines are published in this book series.

With this book, and the series as a whole, we wish to initiate a dialogue with (and among) you and other modelers. We invite you to share with us your ideas, suggestions, and criticisms of the book, its models, and its presentation format. We also encourage you to send us your best STELLA models. We intend to make the best models available to a larger audience, possibly in the form of books, acknowledging you as one of the selected contributors. The models will be chosen based on their simplicity and their application to an interesting phenomenon or real-world problem. Keep in mind that these models are mainly used for educational purposes.

So register now by sending us your name, address, and possibly something about your modeling concerns. Invite your interested colleagues and students to also register with us now. We can build a modeling community only if we know how to make, and maintain, contact with you. We believe that the dynamic modeling enthusiasm, the ecolate skill, spreads by word of mouth, by people in groups of two or three sitting around a computer doing this modeling together, building a new model or reviewing one by another such group. Share your thoughts and insights with us, and through us, with other modelers. Information for writing to us follows:

Bruce Hannon
Jubilee Professor
University of Illinois
220 Davenport Hall, MC 150
607 S. Mathews Avenue, Urbana, IL 61801
Phone: 217 333-0348, Fax: 217 244-1785, Email: bhannon@uiuc.edu

and

Matthias Ruth
Roy F. Weston Chair in Natural Economics
University of Maryland
2101 Van Munching Hall
College Park, MD 20782
Phone: 301 405-6075, Fax: 301 403-4675, Email: mruth1@umd.edu

Index